# DOD SCIENCE AND TECHNOLOGY

## *Strategy for the Post-Cold War Era*

# DOD SCIENCE AND TECHNOLOGY

## *Strategy for the Post-Cold War Era*

J. DOUGLAS BEASON

An Industrial College of the Armed Forces Study

NDU PRESS

NATIONAL DEFENSE UNIVERSITY PRESS
Washington, DC

## National Defense University Press Publications

To increase general knowledge and inform discussion, the Institute for National Strategic Studies, through its publication arm the NDU Press, publishes *Strategic Forums*; McNair Papers; proceedings of University- and Institute-sponsored symposia; books relating to U.S. national security, especially to issues of joint, combined, or coalition warfare, peacekeeping operations, and national strategy; and a variety of other works designed to circulate contemporary comment and offer alternatives to current policy. The Press occasionally publishes out-of-print defense classics, historical works, and other especially timely or distinguished writing on national security.



---

NDU Press publications are sold by the U.S. Government Printing Office. For ordering information, call (202) 512-1800 or write to the Superintendent of Documents, U.S. Government Printing Office, Washington, DC 20402.

---

First Printing, August 1997

**ISBN 1-57906-045-5**

# CONTENTS

# ACKNOWLEDGMENTS

This book could not have been written without the help of my committee at ICAF and the many people who allowed me access to their valuable time. The most important and pivotal role was played by the head of my committee, Col (Dr.) Mary McCully, who not only read through many drafts, but kept me going back and insisting that I say what I really mean. I'd also like to acknowledge the other NDU faculty members on my committee, Dr. Joe Goldberg, Dr. Marty Libicki, and Dr. Rita Wells for directing and helping me to shape this study. Many thanks are also given to those who allowed me time interviewing them, bending their ear, and giving me insight: The Honorable Dr. Karl Erb, The Honorable Gil Decker, The Honorable Skip Johns, MGen Dick Paul, Ms. Michelle Van Cleave, Mr. George Singley, Dr. Finner Milton, Dr. Tom Killian, Col Rich Davis, Dr. Brendan Godfrey, Dr. Ed Feigenbaum, Dr. Gene McCall, Dr. George Abrahamson, Dr. Bob Selden, Dr. Mitch Nicolitch, Dr. Tom McCann, and Dr. Greg Canavan.

# DOD SCIENCE AND TECHNOLOGY

## *Strategy for the Post-Cold War Era*

# 1. THE PAYOFF OF LONG-RANGE R&D

A hiatus exists between the inventor who knows what they could invent, if they only knew what was wanted, and the soldiers who know, or ought to know, what they want and would ask for it if they only knew how much science could do for them. You have never really bridged that gap yet.

Winston S. Churchill
*The Great War*

Investments in science and technology are critical to military preparedness, enabling us to stay at the cutting edge of new developments so that our Armed Forces remained the best trained, best equipped, and best prepared in the world.

Our strategy of investment and international cooperation in science and technology will better assure our success today and in the future.

President Bill Clinton
*National Security Science and Technology Strategy*

The Gulf War fought by the United States-led coalition forces in 1991 was a remarkable display of firepower and bravery. The Iraqi military was relentlessly pounded by airpower for 4 weeks before the 100 hour ground war brought the devastated and crushed enemy to their knees. The Gulf War has been called the "First Space War" by some, and the "First War in the New World Order" by others.[1] Even more laudatory, it has been hailed for showcasing the readiness, the training and morale of America's fighting men and women. And there are many remarkable feats of bravery chronicled that contributed to the decisive victory of the coalition forces.

However, as much as bravery, training, and morale contributed to this overwhelming victory, it would be a slap in the face of every patriotic veteran to say that the men and women of the 1990s displayed more professionalism than the soldier-citizens of our earlier wars. The quick reaction of fighter pilots, the coolness of head and situation awareness shown by both the soldier and naval personnel—these were not due to biophysical advances that had evolved over decades of fighting. Certainly they were important in the Gulf War, but these reactions were also important in earlier conflicts, and they have not increased substantially over the years, despite the best training in the world. To be blunt, our airmen, our soldiers, and our seamen are good, even great—but they are not *that* much better, either mentally or physically, than the veterans who preceded them in the past.

So then, what was it that made such a decisive victory possible? What was the miraculous stuff that enabled the U.S.-led forces to flatten and destroy Saddam Hussein's military like a steam roller running amok in a bed of flowers? To pound him from the sky, relentlessly at will? And in any weather, at any time of day, and with astonishingly accurate precision?

The obvious answer was the equipment. The tactics and training certainly contributed—the 1975 Israeli-Arab war showed how tactics and training can provide the winning edge. And General Schwarzkopf's famous "Hail Mary" offensive surely contributed to a quick, decisive victory in the Gulf War. But the telling difference in *Desert Storm* was not the tactics, the American psyche, morale, or even machismo. It was a myriad of things, but with one common underlying thread: the breathtaking accuracy of precision weapons, dropped by all-weather fighters in the dead of night and guided by infrared sensors; the Tomahawk missiles launched from Navy ships; the awesome firepower and agility of American tanks; the mindboggling advances from exploiting space: GPS navigation, communications and remote sensing. This was the stuff that overwhelmed the Iraqi army, and it was all due to *scientific advances* and the resulting *technology* that allowed these weapon systems to be produced.

The U.S. military's overwhelming technological advantage of its weapons systems is directly due to research investments made 15 to 20 years ago.[2] This well-substantiated claim includes such examples as the F-117 stealth fighter, the B-2 stealth bomber, cruise missiles, lasers, micro-electronics, submarine and land platforms, and space assets to include a few.[3] A direct link can be traced from scientist to warfighter, and the power of keeping a finely honed technological edge has inescapably been proven. It pays off to invest in science and technology, and America has reaped the benefits.

But the barbarians are at the gate, ready to storm the city, intoxicated with budgetary fever, downsizing, and a charge to exploit the commercial marketplace. Some of these strategies are necessary and, if applied prudently, will strengthen our national defense. As with all good strategies, however, there are some implementors who take their tasks to the extreme. The seed corn of tomorrow's weapons, the R&D investment of today, is in danger of being starved before it can germinate, lay root and sprout to feed tomorrow's military. And all in the name of "focusing on the warfighter."

This is not an all-encompassing view. The long-range benefits of basic research are not lost on U.S. National Security Science and Technology Policy, written by the National Science and Technology Council and fully endorsed by President Clinton:

> Research sometimes pays immediate dividends, with a transition directly from laboratory bench to defense systems in the field. But most often the full impact of research is not apparent until much later. It is only in hindsight that the patterns of research which spawned revolutionary military capabilities—radar, digital computers, semiconductor electronics, lasers, fiber optics, and

> navigation systems capable of great accuracy—are discernible. Thus, in planning our research programs, we focus not only on immediate needs but also on opportunities that will sustain our technological edge far into the future.[4]

Winning the war is the whole purpose of having a Department of Defense: To ensure for the continuing national security of the United States. But preparing for the *next* war is also a priority, and the crime of not preparing for today's war is just as heinous as forcing our children to fight the *next* war with today's technology simply because of a non-wavering focus on today.

Those in the present do not have a lock on shortsighted vision. Circa 1800, Napoleon lectured Robert Fuller on the futility of powered ships: "What, sir? Would you make a ship sail against the wind and currents by lighting a bonfire under her deck? I pray you excuse me. I have no time to listen to such nonsense."[5] In 1930 the eminent physicist Ernest Rutherford exclaimed, "The energy produced by the breaking down of the atom is a very poor kind of thing. Anyone who expects a source of power from the transformation of these atoms is talking moonshine."[6]

Some people exposed to visions of the future may be too tied to the past, such as Secretary of War Newton Baker, who, when told of Billy Mitchell's claim that airplanes could sink battleships, growled, "That idea is so damned nonsensical and impossible that I'm willing to stand on the bridge of a battleship while that nitwit tries to hit it from the air." Then there was Major General John K. Herr who said in 1938, "We must not be misled to our own detriment to assume that the untried machine can displace the tried and proven horse."[7] Or as late as 1939, when Rear Admiral Clark Woodward sniffed, "As far as sinking a ship with a bomb is concerned, it just can't be done."[8]

It is easy to concentrate on today while putting off tomorrow, but the lessons of the past show that we would not be where were are today if it were not for forefathers' vision. We cannot ignore the future, a lesson that must be learned every generation. Jacques S. Gansler eloquently summarizes this view:

> To get the maximum for its R&D dollars, DOD will have to stop pushing traditional weapon systems and will have to focus more on "breakthrough" technologies that can make a qualitative difference in military advantages—for example, inexpensive, "smart," non-nuclear weapon systems ... At present there is little institutional support for less traditional systems. Special organizations will have to be created to work on such systems.[9]

With the collapse of the former Soviet Union and the subsequent end of the Cold War, there is some talk that the United States will not have to face any formidable enemies for at least two decades.[10] It is generally believed that Third World countries do not have the necessary infrastructure to mount a credible threat. However, in April 1990, Mu'ammar Qaddafi, speaking to fellow Libyans, thundered, "Did not the Americans almost hit you yesterday when you were asleep in your homes? If they know you have a deterrent force capable of hitting the United States they would not be able to hit you. Because if we had possessed a deterrent—missiles that could reach New York—we would have hit it in the same moment. Consequently we should build this force."[11]

Our adversaries are not as content as we are to stumble toward the future. Even today, 5 years after surrendering to the coalition forces, Iraq is vigorously trying to keep its high-tech edge in missile technology and biological and chemical warfare.[12] This is clearly an advanced development effort having a focused military use. Furthermore, reports appear weekly in the mass media of weapon proliferation to Third World countries.

Technology made the difference in our military strength and got us to where we are today, but technology did not miraculously appear. It is grounded in a foundation of science, creatively generated by the long-range research efforts of thousands of people over decades. It is this foundation for technology that this book addresses, and that foundation is long-range research in the basic sciences. Merely supporting long-range research won't hack it; it has to be done smartly, with common sense.

This book gives insight to the history behind long-range research, the rationale for it, the structure it has taken, and some recommendations for using common sense while keeping it going. For if there is any one area that our generation can improve to ensure our children's freedom, it is by investing in the future by (sanely) supporting long-term research in the basic sciences.

## NOTES

1. General Donald Kutyna (USAF ret.). In a speech before the American Defense Preparedness Association's C4I Symposium "Employing Space for Joint Warfighting Success" on March 28, 1996, disagreed that the Gulf War was indeed the "First Space War." General Kutyna strongly pointed out that the "real" first space war will involve actual offensive or defensive weapons in space; the Gulf War was merely the first war where space played an important and visible role.

2. Genevieve J. Knezo, "Defense Basic Research Priorities: Funding and Policy Issues" (Washington, DC: Congressional Research Institute, 90-506 SPR, October 24, 1990), 12.

3. Institute for Defense Analysis, "Report of the Task Force for Improved Coordination of Science and Technology Programs" (Washington, DC: August 1988), 6.

4. Executive Office of the President, Office of Science and Technology Policy, National Security Science and Technology Strategy, (Washington, DC: U.S. Government Printing Office, 1995), 16.

5. Kenneth L. Adelman and Norman R. Augustine, *The Defense Revolution: Intelligent Downsizing of America's Military* (San Francisco, CA: Institute for Contemporary Studies Press, 1990), 66.

6. Ibid.

7. Ibid., 65.

8. Ibid.

9. Jacques S. Gansler, *Affording Defense* (Cambridge, MA: MIT Press, 1989), 235.

10. I gave a presentation at the White House's Office of Science and Technology Policy (OSTP) in July, 1994, during which several OSTP staffers (and a senior administration official) were present. At the end of the presentation, it was unanimously echoed that the consensus of the staff was that no credible military force will be encountered by the United States for the next 20 years.

11. Adelman and Augustine, 62-63

12. *The Washington Post*, 22 March 1996, reports on new findings in Iraq of hidden chemical and biological warheads being clandestinely shuttled from one military installation to another in hopes of evading United Nations inspectors.

# 2. INVESTING IN THE FUTURE

> Our nation's security derives from a combination of diplomatic leadership, economic strength and military might. Advances in science and technology underlie this strength, giving rise to the discoveries that lead to new industries and to the improvements that make our industries more efficient and environmentally sound. By engaging economies abroad, cooperation in science and technology integrates into a larger economic and political order that acts against division and conflict.
>
> President Bill Clinton

For the first time in nearly 50 years our nation has not been at war, and for the past 50 years our nation's military might has been fueled by an ever-increasing growth of technological achievements. But these achievements have fueled more than our military might: Research and Development fuels *business*, the very engine that drives the U.S. economy, and a pillar of our national power.[1]

## THE FRUITS OF RESEARCH AND DEVELOPMENT

Since the beginning of World War Two we have seen the introduction of precision-guided weapons, the atomic bomb, the ballistic missile, the laser, computers, jet aircraft, satellites, radar—the list of scientific and technical achievements is staggering. The pace of technology is also increasing, along with the number of those who are making the advances. Although scientists and engineers performing R&D comprise less than 1 percent of the total U.S. labor force, the rate of growth of these professionals outstrips any other profession, rising from 55 R&D scientists and engineers per 10,000 in the labor force in the mid-1970s to 76 per 10,000 in 1989.[2]

In the millennia since records have been kept, it is estimated that the world has seen a doubling—a 100 percent growth—in knowledge from the dawn of time until the 1950s, and that knowledge has since doubled *again* since the 1950s.

The growth of knowledge has spilled over to the warfighter. Today's warrior is now fighting with more technologically sophisticated weapons than in the past, and this has resulted in fewer warriors being needed to fight on the battlefield. Technological advances in warfare have been a double-edged sword, however, for while the number density (the number per square kilometer) of combatants may have decreased throughout the years, firepower has increased. Figure 1 shows the dramatic decrease in number density made possible by the introduction of then state-of-the-art weapons. The firepower increase may be understood by considering the way that technology has enabled few warfighters to levy more damage: the range of a spear was extended by the bow and arrow; that range was extended by a bullet; that range was extended even farther by an artillery shell; and that range was yet extended by missile technology.

## INCREASING MILITARY EFFECTIVENESS THROUGH SCIENCE AND TECHNOLOGY

In 1945, J. F. C. Fowler enumerated five qualitative parameters to characterize the power of a weapon, of which he gave the "range of action" as the highest priority:[3]

- Range of Action
- Striking Power
- Accuracy of Aim
- Volume of Fire
- Portability.

**Figure 1:** ***Manpower Density on the Battlefield (per square kilometer)***[4]

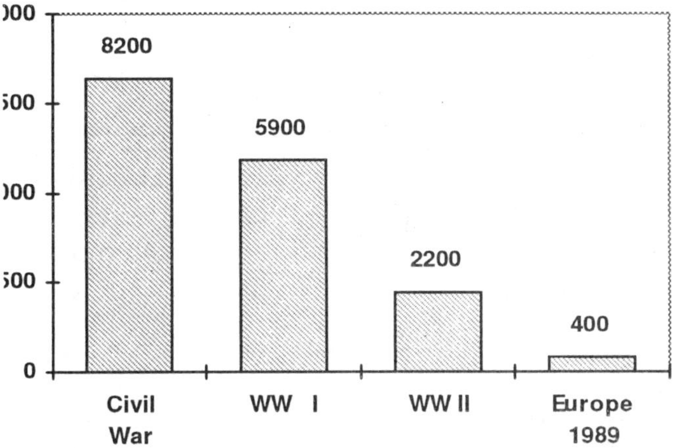


Colonel Simon P. Worden expanded on this concept by deriving the "military effectiveness" as a basic measure of a weapon's military power.[5] Effectiveness may be defined in terms of the "brightness" (a term frequently used by laser engineers to measure the capability of a laser) per unit time, or rather the measure of a weapon's range, accuracy, and power per unit time, all rolled into a single number. Weapon effectiveness is listed in table 1 and is illustrated in figure 2.

Note that the military effectiveness is presented in a compact form as an exponential number—meaning, of course, that bullets have a "military effectiveness" a *hundred times* greater than arrows, and that ICBMs are $10^4$, or 10,000 times more effective than artillery in 1900. Although military tactics and strategy have played a role in increasing the effectiveness of these weapons, the advances in military effectiveness are chiefly due to one reason: the exploitation of science and technology. Figure 2 shows the dramatic increase in military effectiveness on a logarithmic scale; that is, the y-axis is shown as exponential powers of 10, so that the maximum value of "25" is not a simple factor of 5 greater than "20," but rather $10^5$, or *100,000 times greater*.

**Table 1. *Weapon Effectiveness*[6]**

| Era (year AD) | Weapon | Time[a] | Brightness (Joule/Sr) | Firing Rate (per sec) | Effective-ness (Joule/Sr/sec) |
|---|---|---|---|---|---|
| 1000 | Arrows | 6 Months | $10^8$ | $10^{-2}$ | $10^6$ |
| 1500 | Bullets | 3 Months | $10^9$ | $10^{-1}$ | $10^8$ |
| 1800 | Artillery | 1 Month | $10^{12}$ | $10^{-1}$ | $10^{11}$ |
| 1900 | Artillery | 1 Week | $10^{14}$ | 10 | $10^{14}$ |
| 1930 | Aircraft | 1 Day | $10^{19}$ | $10^{-1}$ | $10^{18}$ |
| 1950 | Aircraft | 1 Day | $10^{23}$ | $10^{-2}$ | $10^{21}$ |
| 1970 | ICBM | 1 Hour | $10^{23}$ | $10^{-1}$ | $10^{22}$ |
| 2000 | SBKKV | 1 Hour | $10^{23}$ | 10 | $10^{23}$ |
| 2020 | Laser | 5 Min | $10^{22}$ | $10^2$ | $10^{24}$ |

[a]"Time" refers to both the time period of battle and the time it takes to get into position to engage the weapons.

The technology present in the battlefield keeps growing. Tomorrow's battlefield will consist of global networks keeping track of targets, sophisticated sensors, combatants, platforms (including stealth), and long-range, conventional (non-nuclear), high-precision, extremely accurate weapon systems (both manned and unmanned)—all linked with digital computers.[7]

The point of this exercise is to show that there is an *exponential increase of military effectiveness due to advancements science and technology*. Not just increases in 10 percent, or even a doubling of effectiveness. But true factors of *many thousands of times, all due to science and technology.*

Figure 2. *Increase of "military effectiveness (log scale)*

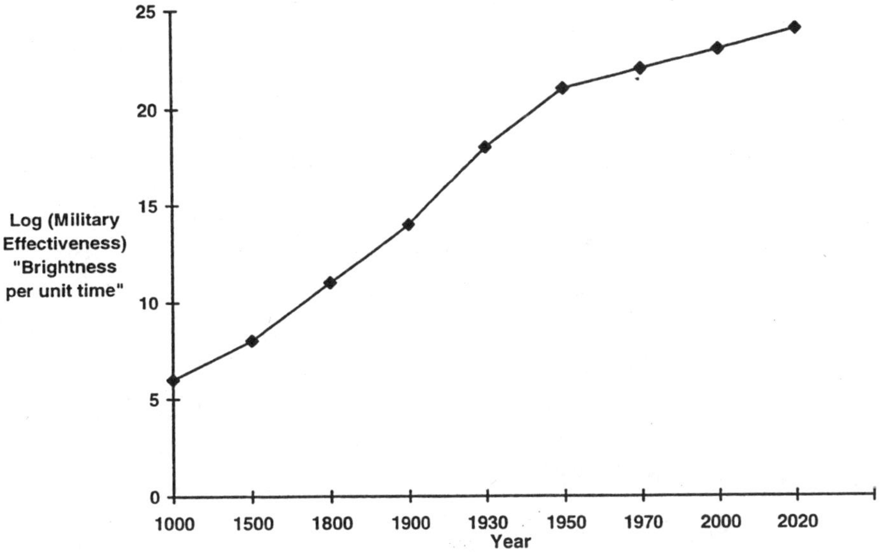


Source: Simon P. Worden, *SDI and the Alternatives*(Washington, DC: National Defense University Press, 1993).

The exploitation of science and technology (S&T) is thus a critical factor in producing the next generation of weapons. Advances in S&T do not happen overnight, however, nor do they happen in a vacuum. In the words of a former researcher at the prestigious Bell Laboratories, "Quality work requires sustained support. You just can't turn on the spigot and have Nobel Prizes overnight."[8] So as in any other endeavor, S&T needs to be nurtured, looked after, and sustained, or it will die.

## TRENDS IN RESEARCH AND DEVELOPMENT

Research and Development (R&D) in the United States has typically been funded at greater levels than any other country. Approximately $176 billion was spent in industry, government, and universities in 1994, or 2.6 percent of the GDP.[9] In 1992 (the last year that comparative data were available), the United States spent over 28 percent more on R&D than Japan, Germany, and France *combined*.[10]

The majority of these funds was spent on defense-related R&D, and the fruits of this effort are borne in the quality of American military technology. The defense share of federal R&D funding reached a peak in 1987 during the Reagan buildup, accounting for 69 percent of the total federal R&D budget. Between 1992 and 1994 the percentage of defense-related R&D slipped from 59 percent to 56 percent—in line with the Clinton administration's goal of producing an even split between defense and nondefense R&D funding.[11]

But an even split in federal R&D funding does not imply an even split between Federal Government and industry *performing* defense R&D (e.g., defense laboratories versus industry laboratories). The nation's industry is already responsible for the majority of R&D spending in the United States, accounting for $125 billion in 1994, or 71 percent of the national total. Although this R&D spending in industry is highly concentrated—eight industrial sectors account for more than 80 percent of the total industry R&D spending, with the aircraft and communications industry in the top two positions—claims by the present administration that these sectors perform a majority of their R&D in defense-related areas are simply not substantiated.[12]

In addition, industrial firms are continuing to increase their support of university R&D, growing from 3.9 percent of the total funding that universities received in 1980 to 7.3 percent in 1993.

Overall, university R&D spending has increased as well, rising from 9 percent of the nation's total in 1985 to over 12 percent in 1992.[13]

With the end of the Cold War, the present administration has recognized that maintaining a separate defense and commercial industrial base is not efficient. The Department of Defense is now working more closely with industrial firms engaged in commercial and dual-use production, including performing dual-use R&D. It is the administration's goal not only to strengthen the civilian industry, but also to promote "the cost-effective development of new technologies for national defense and stimulate the creation of an integrated civilian-military industrial base."[14]

However, the administration has realized that commercial technology simply cannot fill all the military's needs. Tanks, fighter aircraft, nuclear submarines, new stealth platforms, and many other applications require defense-unique technologies. These technologies must be grown through a variety of mechanisms, from a combination of industrial, university, defense and "national" (e.g. Department of Energy, NASA, and other federal agency) laboratories. But for all the infrastructure set up to transition technology to defense, without a "seed" of long-range research, the germination of growth that allows for creative ideas to evolve into new technology, no advances will be made. We will be forever stuck with merely optimizing what is known—a slow, evolutionary increase in efficiency rather than in the promise of explosive growth (figure 2).

The present administration recognizes this and has gone on record by stating,

> Some analysts . . . [say] that the Nation is spending too little on *basic* research that will drive tomorrow's revolutionary breakthroughs. This concern is supported by empirical evidence that suggests there are large unexploited economic gains to be realized from raising our society's level of scientific activity and

> technological research and development; in the past, the social rate of return on such investments has been high. As a central component and stimulus of U.S. innovation, Federal R&D investment can lead technical innovation nationwide and affect the Nation's military posture, a variety of social objectives, and the competitive performance of U.S.-based firms in domestic and foreign markets.[15]

R&D has an impressive payoff history, a propensity of giving back more than what is invested. It is generally accepted that the "Return On Investment" of long-range research will reap large rewards in the future. Onthe average, R&D investment in the United States has exceeded the rate of return available in private industry, performing at levels as high as 423 percent (optical fibers) to an industrywide rate of return between 30 and 80 percent.[16]

But if the rates of return of long-range research have remained so consistently high, then why is not more private capital put toward exploiting it, and why isn't the Federal Government championing it even more than it is? The quintessential reason is that long-range research is *risky*. There is no guarantee that a winner will be produced for every research project started—but yet, the returns have the chance of being "sky high." An analogy is buying a lottery ticket. The chances of winning are extremely small, and it has a near "infinite" payoff. But no one knows exactly which lottery ticket will be the lucky one. Unlike the relatively low cost of buying a lottery ticket, funding long-range research costs *a lot* of money—up to billions of dollars a year. To make things worse, the marginal return for long-range research is not linear—in fact, there is no assurance of making a return on the investment at all.

The purpose of this book is to stress the importance that long-range research has to the strength of our nation and the need for a continued strong investment in the future. In particular, basic research and applied research—the generally accepted constituents of long-range research—have a place in our nation's defense but this fact is often ignored. This could very well prove to weaken our country.

It is with this fact in mind—that long-range research plays a fundamental part as the foundation for our nation's strength—that a case for long-range research will be presented.

## NOTES

1. Jacques S. Gansler, *Affording Defense* (Cambridge, MA: MIT Press, 1989), 169.

2. National Science Board, NSB 93-1, "Science and Engineering Indicators" (Washington, DC: U.S. Government Printing Office, 1993), p. 67.

3. J. F. C. Fuller, *Armament and History* ( New York: Scribners, 1945), 7.

4. Kenneth L. Adelman and Norman R. Augustine, *The Defense Revolution: Intelligent Downsizing of America's Military* (San Francisco, CA: Institute for Contemporary Studies Press, 1990), 55.

5. Simon P. Worden, SDI and the Alternatives, (Washington, DC: National Defense University Press, 1991), 13-15.

6. Ibid., 14.

7. Adelman and Augustine, 53.

8. Quoted from Dr. Cammy Abernathy, currently a Professor of Material Science at the University of Florida, Gainesville, in an article by Robert F. Service, "Relaunching Bell Labs," *Science* 272, 3 May 96, 639.

9. *Economic Report of the President* (Washington, DC: U.S. Government Printing Office, 1995), 163.

10. Ibid.

11. Ibid.

12. Ibid. The report claims that 76 percent of the federal money that goes to support R&D in aerospace and communication firms (which is one-fourth of the total funding that private industry spends to support its R&D) is used "primarily for development of weapons and related systems of military application." The report gives no breakdown of these funds, (that is, as to what military systems are supported) and does not distinguish between it's own definition of dual-use: Does R&D performed by industry on flat panel displays help the military (where only a limited number of F-22 fighters may use the display), or does it actually help the commercial marketplace, as industry wrestles with taking back the lead from Japan? My point here is that the Administration cannot have it both ways—they may decry the amount of R&D

funding spent on defense R&D, yet it may well be fulfilling their own goal of accomplishing dual-use R&D. They do no one a favor by muddying the water with unsubstantiated claims that may have other purposes in mind.

13. Ibid., 164.
14. Ibid., 164-165
15. Ibid., 164.
16. Ibid., 122.

# 3. THE EVOLUTION OF DEFENSE RESEARCH

*In science, the few must dictate opinion to the many.*

Edward Bulwer Lytton

The National Science Foundation (NSF) was established by the NSF Act of 1950. It was founded to support and initiate basic scientific research in the United States in the environmental sciences, physical sciences, life sciences, social sciences, mathematics and engineering. The NSF is charged with funding, and therefore strengthening, the nonmilitary basic research base in the United States. The NSF defines research as:

> The systematic, intensive study directed toward fuller scientific knowledge or understanding of the subject studied. Research may be classified as either basic or applied. In *basic* research the investigator is concerned primarily with gaining a fuller knowledge or understanding of the subject under study. In *applied* research, the investigator is primarily interested in the practical use for the purpose of meeting a required need.[1]

Many similarities exist between nondefense and defense-related research; some major differences will be explored later in this chapter. However, before the evolution of defense-related research is examined, it is necessary to first understand the definitions in research, and then how research in general contributes to society.

## DEFINITIONS IN RESEARCH

### BASIC RESEARCH

Basic research is considered long-range, nonfocused inquiry that advances the state-of-the-art frontiers of fundamental knowledge.[2] Basic research may never have a practical application and is directed toward solving the axiomatic problems of nature. Although basic research may never have a practical application, seemingly insignificant discoveries have resulted in world shaking phenomena that have profoundly affected human affairs. Who would have guessed that the discovery of beta decay, or the benign scattering of atomic particles, would have ultimately led to the terrible destructive forces that are found in the atomic bomb? Or that the discovery of an electronic "gate" known as the first transistor in solid state physics would someday result in the computer chip and the dawning of the "information era"?

**Linear Versus Non-Linear Investment.** Basic research involves a long-term, risky investment with no assurance of success—but with the intoxicatingly possibility of having an extremely high payoff. Since it is impossible to exactly predict the future and know which discovery will result in the next "silver bullet" to solve the current high-priority problem of the day, an investment must be made in *many* areas with the hope that a discovery in one of the fields will produce such a winner.

Clearly DOD or the nation does not have the money nor the people to invest in all areas of science. In 1995 the nation's total investment in conducting Research and Development was on the order of $171 billion a year, with the Federal Government funding $60.7 billion; and of that amount, DOD invested $35.1 billion in Research, Development, Test and Evaluation (RDT&E).[3] But as much money as this is, we can't pursue all areas and therefore must choose wisely where to invest.

Therefore, the question then becomes in *which* areas should the nation invest? Should one peer through a "soda straw" and hope that by funding a few specific areas in science, that a world-changing discovery will occur? This assumes that practical technology leaps directly from advances made in long-term research, a "linear" view of investment.

But looking back on revolutionary technologies, rarely does one find a direct, linear link that leads from the initial scientific discovery to the product itself. Rather the path is quite circuitous, often jumping fields (i.e. cross feeding from unrelated areas) and being refined by advances in other areas of science. This is called having an *interdisciplinary* approach to research—using advances in one field to advance the state-of-the-art in an unrelated field.

A good example of "non-linear" scientific evolution may be found in the invention of the fiber optic cable, now a highly efficient means of transmitting data. In the early 1900s, Albert Einstein published his results on stimulated emission, and from his theory were derived some parameters known as the "Einstein coefficients." Advances in the fields of atomic physics, mathematics, optics, and electronics continued to made, building on the Einstein coefficients, until the microwave laser, or the "maser," was discovered in the 1950s. Much research was accomplished trying discover a way to extend the physics to inventing an optical maser (laser), when it was realized that simply by placing optical mirrors on either side of a laser amplifier, the laser would work. These advances were all ripe to add to the next discovery: Combining the fields of laser physics, optics (internal reflection), electromagnetism (standing wave modes), and quantum theory (stimulated emission), would contribute to the invention of the fiber optic cable—a true cross-cutting of separate fundamental disciplines, a nonlinear evolution.

This shows but one example of how a highly useful application of technology did not appear overnight. More importantly, it shows how it was drawn from disparate fields.

It was not discovered from a soda straw view of the world, that is, setting a requirement and expecting that an investment in a few fields of science will produce a world-changing discovery. That is the danger of assuming long-term research will always produce practical results. It just may, but probably not in the way anyone envisioned.

Thus, a danger exists if management demands a technological invention by forcing it to spring from a linear path, from scientific discovery straight to technological application. In reality, nonlinear, *cross-feeding* discoveries from diverse fields will be the most likely way a scientific breakthrough is accomplished. Certainly, following linear paths is more satisfactory to management than pursuing opaque interdisciplinary research: linear thinking can be understood, by political scientist and businessman alike. But *banging on one area in basic research will not necessarily produce an advanced technology "silver bullet" in that same area.*

**DOD Research.** As is every other organization, DOD is constrained by money and people, and it must choose wisely in which areas to invest. Universities perform approximately half the basic research used in DOD, with scientists and engineers at DOD laboratories, along with industry, nonprofit research institutions, and other federal laboratories performing the rest of the research.[4] Care is taken to inject diversity into the areas that are funded; as a result, the DOD basic research program is structured around 12 "fields of inquiry:"[5]

- Atmospheric and Space Sciences
- Biological and Medical Sciences
- Chemistry

- Cognitive and Neural Sciences
- Computer Science
- Electronics
- Materials Science
- Mathematics
- Mechanics
- Ocean Sciences
- Physics
- Terrestrial Science

These fields provide the basis for basic research in DOD. The next step is transistioning from this research that has no focused objective to one that can be channeled toward a military purpose.

### APPLIED RESEARCH (EXPLORATORY DEVELOPMENT)

Applied research is differentiated from basic research in that it has a primary focus toward specific objectives, either as a final product or process. This objective may be to serve a commercial need or to satisfy a requirement in a much larger, more sophisticated system.[6] Applied research may be long range, or shortrange, as long as the specific objectives are met. For example, heat-resistant shielding for ballistic missile warheads was developed out of applied research efforts. Focused research in ablation physics, material science, and chemistry, coupled with techniques refined in mechanical engineering, made possible the development of tape-wrapped carbonic phenolic (TWCP), a primary heat-resistant shield.

The Department of Defense uses the term "exploratory development" interchangeably with applied research. The main goal of exploratory development is to bridge the gap between basic research (that having no practical application) with that of the initial development of hardware for use in a system. It provides the "proof of concept" experiments that take fundamental scientific discoveries off of the "laboratory bench" and shows that they have a practical (although not necessarily yet a military) application.

The Department of Defense has identified 19 technology areas that will provide the conduit for transitioning from exploratory development to the next phase of the research and development process, advanced technology development:[7]

- Aerospace Propulsion and Power
- Air vehicles and Space Vehicles
- Battlespace Environments
- Biomedical Applications
- Chemical and Biological Defense
- Individual Survivability and Sustainability
- Command, Control, and Communications
- Computing Software
- Conventional Weapons
- Electronics
- Electronic Warfare and Directed Energy Weapons
- Environmental Quality and Civil Engineering
- Human Systems Interface
- Manufacturing Science and Technology
- Manpower, Personnel, and Training
- Materials, Processes, and Structures
- Modeling and Simulation
- Sensors
- Surface/Under Surface Vehicles and Ground Vehicles.

## ADVANCED DEVELOPMENT

This phase that incorporates the initial development of hardware for use in a system is called "advanced development" in DOD. Advanced development is structured to directly apply scientific and technological advances to military uses. For all practical purposes, the exact differences among basic research, applied research (exploratory development), and advanced development lie in the *funding*, the budgeting and accounting of the research, rather than the execution. There are no "hard and fast rules" as to when the boundaries of basic research end and applied research (exploratory development) begins.[8] However, long-range research is generally considered to consist of basic research and exploratory development; short-range research typically means the advanced development phase.

Basic research, exploratory development and advanced development are all tightly coupled—an advanced development concept cannot be accomplished unless the practical applications of the science are discovered in the exploratory development stage. In turn, the practical application of a scientific discovery cannot be assessed until the discovery is made.

This results in one of the most perplexing situations the military has to face: how to fund the correct broad-based area of basic science in order to obtain specific solutions to its war-fighting requirements. This is the true crux of the issue. The methodologies used to attack this problem are studied later in this book.

### SCIENCE AND TECHNOLOGY (S&T)

In the Department of Defense, the general field of science and technology (S&T) is defined as a combination of basic research, exploratory development (applied research), and advanced development. S&T then may be viewed as a dynamic, ever-advancing process whereby discoveries made in basic research are fed into a "base" data bank of knowledge for applied research to exploit; in turn, once the scientific discovery is shown to have some practical application via the exploratory development phase, the military application is tested through the advanced development process. Applied research and advanced development stand on the shoulders of basic research to accomplish their goals in solving a particular problem.

### RESEARCH AND DEVELOPMENT (R&D)

Research and development (R&D) encompass all science and technology and, further yet, *develop* S&T in a systematic manner toward useful products, processes, materials, or systems. Thus, R&D is a superset of S&T.

In a defense-related example, S&T may be performed to invent new air-to-air missile technology, such as discovering new search algorithms or inventing more energetic and stable missile fuel. The R&D program associated with developing this new missile includes the testing, evaluation, and assessment of missile performance, including the cost of fuel that any carrier airplane may use in conducting these tests. Therefore, R&D is a much broader definition of the technological process, pulling in all the supporting infrastructure and administrative support necessary for achieving the S&T goals.

## INDUSTRIAL RESEARCH

The roots of defense-related research are grounded in the history of industrial research laboratories. Understanding today's defense research needs and infrastructure directly springs from knowing how industrial research laboratories evolved in the United States.

### THE CONNECTION BETWEEN INDUSTRY AND SCIENCE

Science is crucial to the health and conduct of modern industry. It is the foundation for technology and the fuel that powers the engine of change that advances our economy. Without advancement in science, industry would eventually grind to a halt, stagnating, as technology would no longer provide the impetus for growth. Science begets knowledge, and from understanding and innovation springs economic prosperity.[9] The next section will show some examples of technology that grew from advances in science.

**Science in the Mid-1800s.** Science and industry have not always been as tightly intertwined as they are today. Until the 1870s, advances in science largely evolved separately from industrial technology, neither influencing the direction of industry nor significantly contributing to direct technological breakthroughs.[10] This is especially true of advances in the basic sciences, defined as physics, chemistry, biology, and applied mathematics. For example, the explanation given for natural phenomena, such as the motion of the planets around our sun, did not directly contribute to the growth of the mining, transportation, construction or manufacturing industry.

This rang true for nearly all discoveries in basic science. Rarely did unveiling a fundamental truth result in a breakthrough, or even an appreciation, for what the discovery had to offer. In fact, even today the sirens of waste still wail at the worthlessness generated by those performing research in the basic sciences. In the past, advances in various sectors of the industrial revolution such as shipbuilding, engineering, architecture, mining, and manufacturing were based upon experience of the artisan, rules of thumb, and traditions handed down from generation to generation rather than on advances in science.[11] There was no systematic way of applying the "scientific method" to advancing industrial technology. In fact, the "feel" for industrial technology is still embodied in the phrases "It's more of an art than a science" or "It's like magic."

But the separation of basic science and industrial technology did not last for long. Soon, they became tightly coupled, with technology relying on basic science as its underlying foundation. Advances in technology no longer happened "just by accident."[12] A rigorous transition began to occur, starting from scientific discovery to technological application.

**The Maturation of Technology**. Within a few years of 1875, discoveries in the basic sciences made since the 1600s—an incubation process lasting nearly 270 years—started to feed into the West's industrial technology base.[13] Over this period, the basic sciences laid the ground work for explaining the basis behind the natural sciences, chemistry and physics. These developments culminated in the foundation of rigorous engineering procedures which caused the rapid evolution of technology. Of the hundreds of significant breakthroughs in the basic sciences in the mid-19th century, the following highlights just a few:[14]

*1850s:*

- Clausius' Second Law of Thermodynamics
- Riemann's non-Euclidian geometry
- Maxwell's Kinetic Theory of Gases

*1860s:*

- Dmitri Mendeleev's Periodic Table of the Elements
- Kirchoff's Black Body Radiation
- Maxwell's Electromagnetic Equations

*1870s:*

- Louis Pasteur's work in food spoilage
- van der Waal's Gas Laws
- Gibb's Chemical Thermodynamics

The application of these discoveries was not immediately apparent; however, from out of Maxwell's equations sprang the basis for radio, TV, electronics, and computers; Dmitry Mendeleyev's work on the periodic table established the basis for modern chemistry; Louis Pasteur's work in food spoilage resulted in the science of bacteriology and modern biology; and other foundations for advanced technology were similarly established.

On the surface, it appears that there is a direct connection between discovery and application. Looking in the past, it is relatively easy to show a simple path from creative spark to world-changing technology.

*But the path from these scientific discoveries to the technology used in industry was not direct.* Rather, it was long and circuitous, rarely linear, and never straightforward. One discovery beget another; a new application yielded a springwell of others. Rarely did an application leap directly from the mind of the inventor; it waited instead to be revealed by the technologist as is an onion's inner core—peeled away, layer by layer.

And *that* long and circuitous route is the main theme of this chapter, an idea to be trumpeted by every scientific advocate. It is easy to forget that the labors of scientific research do not quickly bear fruit—the applications of basic research are measured in timescales of decades, and not days. For example, the time between "invention and innovation" for the fluorescent lamp took 79 years; the gyrocompass, 56 years; the cotton picker, 53 years; the zipper, 27 years; the jet engine, 14 years; radar, 13 years; the safety, razor 9 years; and the wireless telephone, 8 years.[15] So although this long time-scale is a drawback of long-range research, the applications have proven that they can change the direction of society.

In the past there existed no formal mechanism for directly transferring either knowledge or advances in the basic sciences to industry. Too often, science was viewed as a topic allowing for interesting digressions, more for conversation than for driving technological change, or in particular, not being able to add to the profit base of companies.

Outside of university-derived research, science served no useful purpose other than existing as a tool for the enlightened person. As such, science was viewed as only that knowledge held by a few old wise men. It was not until the scientists were organized to specifically transfer their knowledge to solving the problems of the times that the real power of scientific research bore fruit.[16] The application of the scientific method—one of rigorous examination and experimentation followed by unbiased replication of data—provided the basis for this organization, the establishment of the research laboratory.

## THE RISE OF INDUSTRIAL RESEARCH LABORATORIES

The first research laboratories in the United States outside of the university setting concentrated on refining techniques for inserting advances in chemistry into industry. In 1836, Charles T. Jackson and James C. Booth established chemical laboratories nearly simultaneously in Boston and Philadelphia.[17] One field that grew out of these endeavors was chemical engineering, now regarded as the foundation for modern chemical technology.

The number of industrial research laboratories in the United States grew from 2 in 1836, to 139 in 1898, and to over 550 by 1918.[18] These laboratories were nearly always associated with large enterprises; the larger the company and more dependent the company is on technology for producing its product, then the greater the likelihood of a research laboratory presence. This may be explained by the invention of mass production and the resulting economies of scale. Although some may argue that large organizations are just too bureaucratic and inflexible to employ research laboratories, their very size is what makes it possible for research laboratories to flourish.

R&D by its very definition is associated with high risk endeavors. In fact, the higher the risk, the higher the potential payoff for the product that the laboratory may develop. R&D laboratories conductong high-risk research, some of which may not pay off for many years, requires investment in extensive facilities and modern equipment and development of a large infrastructure to maintain the R&D laboratory. These large, up-to-date facilities can be created and maintained onlyby large enterprises. Therefore, the presence of a large corporation provides the fertile soil necessary to maintain an R&D laboratory, as well as the necessary funds to provide the start up capital.

Large enterprises often exploit the technology of mass production and economies of scale to maintain their profit. This was particularly true back in the mid 1870s when the number of research laboratories skyrocketed. In order to improve their product they had to make advances in two areas: first, a near-term or "short time-scale" advance in which their product was "tuned" or improved to work better. The second area was a longer term, or "long time-scale" advance, which resulted in a complete redesign of the product or even a new product line. Large companies could afford the necessary costs needed to field an R&D laboratory that feeds innovations into their product line.

This paradigm of performing long-range research in industrial laboratories drove the structure and nature of science and technology for many years in the civilian sector. The industrial in-house laboratories were expected to provide a long-term base, providing an evolutionary—and sometimes, but rarely, a revolutionary—increase in the production line. This continued until the 1980s when American industry realized that it could no longer exploit its commercial success based on the techniques of mass production. Instead, agile manufacturing, or limited, tailor-made research—hitting problems fast and hard with specialized research teams so that they could respond to short-term market pressures—began appearing.[19] Long-range research now focused on the *revolutionary*, rather than the *evolutionary* advances.

The times changed. Industry responded to market pressure and refined the way they conducted business. But throughout the turbulent times, the importance of long-range research remained; it was often sought to provide a revolutionary "silver bullet" to catapult industry ahead of the competition, but mostly it was focused on areas that promised growth.

It is now widely accepted that long-range research in the basic sciences provides the foundation for our nation's technological strength.[20] The Clinton administration has strongly endorsed basic science and has prepared an extensive policy on championing this support:

> America's future demands an expanding knowledge base, which requires investment in our people, institutions, and ideas and cooperation with international partners to expand our access to data and information. Science lies at the heart of that investment—it is an endless, sustainable, and renewable resource with extraordinary dividends. Today's investments in basic science build a foundation for commercial products and services for the future. The nation's commitment to world leadership in science, engineering, and mathematics created the world's leading scientific enterprise, whether measured in terms of discoveries, citations, awards, and prizes, advanced education or contributions to industrial and informational motivation. Our scientific strength is a treasure we must sustain and build on for the future, and this Administration is firmly committed to its support.

The United States has refined a system for selecting excellence in ideas, individuals, and institutions that s extremely competitive and productive. It is a system that achieves quality by emphasizing peer review and promoting creativity. The system cannot always predict the exact areas or nature of scientific breakthroughs or the timeline for fundamental discoveries. Over decades, however, it reliably produces discoveries that enrich the lives and prospects of our citizens and, when transformed to practical cost-effective products, reorganizes old businesses and creates new ones.

The Federal Government has long played a vital role in ensuring American leadership in science, mathematics, and engineering, and investment in basic science continues to be an essential component of our innovation portfolio.[21] It is this policy that provides the blueprint for our nation's investment in long-range research.

## THE ROOTS OF DEFENSE RESEARCH

The need for defense-related research, specifically both applied (or exploratory) development and advanced development, was established in this country's infancy. It was recognized as early as the Revolutionary War, and a precedent was set by President George Washington in his support for defense-related research.[22]

Although the utility of advanced development research in defense was acknowledged in the late 1700s, the path to refining the advanced technology was not yet invented. The concept of conducting basic research, which would feed into a more highly focused applied research stage and then finally move toward advanced technology research, was simply not elucidated. In other words, the role of basic research and its place as a fundamental part of the research process was neither appreciated nor exploited. It was not until the early 1960s, when a fully sanctioned and systematic process was established, that the necessity of conducting basic research was formalized in the Department of Defense.

## SUPPORT FOR BASIC RESEARCH

Although a rigorous process for conducting basic research in DOD was not established until the 1960s, some parts of the military establishment became a champion for supporting basic research early in our nation's history. It was recognized that fundamental technology problems must be overcome in order to field different weapon concepts, and the only way to battle these problems was by the use of a systematic research process. Although the majority of the defense establishment did not acknowledge the importance of this process, nonetheless, "pockets" of support for research sprang up throughout the military. These "pockets" ensured that basic research was accomplished when it was needed, meaning that the lack of support in the rest of the defense community was not as devastating as it could have been.

For example, although the military had no funding specifically authorized for basic research, in 1833 the Navy established the Naval Observatory in Washington, DC. At that time the Naval Observatory advanced research in astronomy and navigation. The Navy was also interested in advancing the state-of-the-art in shipbuilding and was one of the earliest Federal institutions to actively promote basic research.[23]

A quasi-official agency, the National Academy of Sciences, was established on March 3, 1913 to stimulate and encourage the advancement of science in the United States.[24] When it was first established, the Academy was charged with supporting military as well as nonmilitary research. The Academy chartered the National Research Council in 1916 as an official operating arm to specifically promote research related to national defense.

On the recommendation of the Navy Consultant Board, then chaired by Thomas A. Edison, the Naval Research Laboratory was established in 1916. This was the first military organization exclusively devoted to the systematic method of conducting defense-related scientific research. To this day the Naval Research Laboratory remains one of the nation's premier military and nonmilitary research institutions, advancing basic and applied science in fields as diverse as space to high power microwaves.

In 1940 President Roosevelt established the National Defense Research Committee, to "correlate and support scientific research on the mechanisms of warfare" and to "conduct research for the creation and improvement of instrumentality's . . . of warfare."[25]

Although the National Defense Research Committee had no equipment, personnel, or buildings of its own, it was charged with tasking research to government, industry, universities, the National Academy of Sciences, and the National Research Council.

The Office of Scientific Research and Development (OSRD) was established on June 28, 1941, by President Roosevelt to assure "adequate provision for research on scientific and medical problems relating to national defense."[26] OSRD was given power to direct, coordinate and oversee research during the war. The success stories of research conducted under the auspices of OSRD include the development of the atomic bomb, the discovery of antimalarial drugs, the invention of the proximity fuze, radar, and other significant discoveries.

In June 1946, General Dwight D. Eisenhower signed into force an overarching War Department Research and Development policy that would integrate military research with the civilian industrial base. As such, a Joint Research and Development Board was established to act as the coordinating agency for common R&D in the Navy and War departments. This policy was the first formal attempt to consolidate military planning, design and production of weapons using civilian assistance to exploit American industrial and technological capability.[27] However, this move was not viewed favorably by all.

## CRITICS OF DEFENSE R&D

In its September 1946 issue, *Business Week* gave a bleak forecast of what it viewed as the role of defense-related science:

> Partly by design, partly by default, federal support of pure science is today almost completely under military control. Its general direction is being set by military needs; its finances are coming from military funds. The odds are getting better all the time that

> pure scientific research will become, permanently, a branch of the military establishment.[28]

There are still some grumbling that the military presence in science has forever hurt this traditionally bipartisan field—"traditional" only in the sense that the critics do not see the long-standing support of civilian technology by the military. And although critics such as Foerstel keep attacking the perceived problem with such shallow, ill-analyzed "exposes," it is the general consensus that defense-related S&T has attributed to advancing American competitiveness.[29]

Although it is argued that military budgets divert economic resources—capital, personnel, materials, skilled laborers, scientists and engineers—from contributing to nonmilitary production, the nonmilitary sector in fact rides the coattails of military spending. This happens by reinvesting money back into the economy and by exploiting the technical spin-offs. In addition, decreased defense spending, particularly in cutting defense-related R&D, may do little or nothing as far as increasing the U.S. economic growth—especially if the so-called "savings" are not channeled toward productive investment and are instead spent on pork, pet projects, or income redistribution schemes.

## CENTRALIZED CONTROL OF DEFENSE R&D

The National Security Act of 1947 set up the infrastructure for exercising centralized control of defense-related R&D. The founding of the National Military Establishment by this act attempted to rid defense R&D of duplication that existed in laboratories by establishing a Research and Development Board (RDB). The RDB replaced the Joint Research and Development Board and was tasked with the following specific functions:

- Develop an annual defense-wide R&D plan
- Provide guidance to all the military services for implementing the R&D plan
- Continually review and analyze defense R&D facilities
- Make R&D budget recommendations
- Assign one military service the primary responsibility for conducting R&D in specific areas in order to avoid duplication, promote efficiency, or achieve economy.[30]

The RDB operated with committees consisting of both military and civilian personnel, focused on looking at general areas of technology. However, the committees were parochial in nature and gravitated toward parroting their own service's view.[31] Even worse, critics charged that the inherent make-up of the committees resulted in recommendations that advanced the present state of technology by evolution, rather than recommending that truly "high-risk, long term" research be accomplished.[32]

These problems were recognized by 1953 and the RDB was replaced by an Assistant Secretary of Defense for Research and Development. Although the separate military services were responsible for planning, funding and conducting their own R&D, the Assistant Secretary ensured that the military services coordinated their research into an overall coordinated program.[33] However, an Assistant Secretary of Defense for Applications Engineering was also established, and a battle ensued between the two position's responsibilities. The problem was resolved in 1957 when the two positions were combined into the position of Assistant Secretary of Defense for Research and Engineering.

In 1958, DOD formalized research infrastructure solidified into its near-present form when the Department of Defense Reorganization Act of 1958 established the Defense Director of Research and Engineering (DDR&E), making the DDR&E the sixth ranking member of DOD behind the secretary, the deputy secretary and the three service secretaries. The DDR&E became the principal advisor to the Secretary of Defense on basic and applied research, development, test and evaluation of weapons and equipment, design and engineering, and reliability, maintainability, and technology.[34] Further, the DDR&E is also responsible for the direction and control of all defense-related R&D, as well as formulating policy and developing standards.

In the years since the War Department Research and Development policy was first issued, a high was reached in 1968 of over 140 defense "in-house" technical organizations. The buildings and equipment were valued at $4.1 billion and were housed in 10 million square feet of building space on 10.6 hectares of land.[35] With the defense downsizing, resultant base closure and realignment process in action, this number has steadily dropped.

Total outlays for Research & Development and Testing and Evaluation (RDT&E) have continued to increase in DOD, with a high of $37.5 billion reached in 1990. Figure 3 shows the total outlays for DOD RDT&E outlays from 1972 to 1996.[36]

**Figure 3.** ***DOD outlays for RDT&E (billions)***

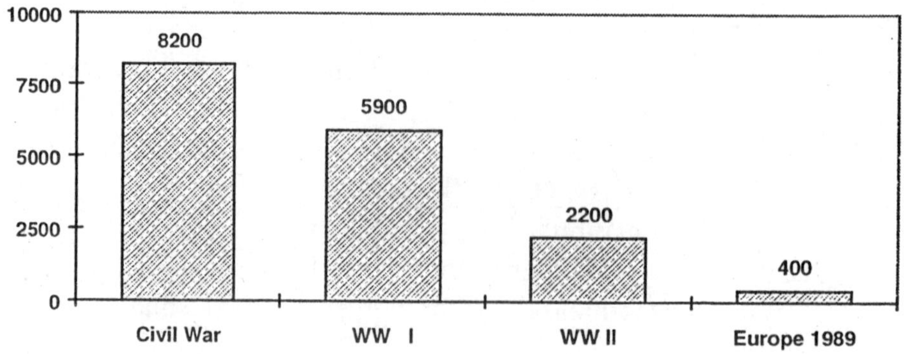

## THE DEFENSE WEAPONS DEVELOPMENT PHASE

The defense weapons acquisition program consists of three phases: concept formulation, contract definition, and acquisition. Research and Development is a major part of the concept formulation phase and encompasses all of science and technology (basic research, applied research or exploratory development), as well as advanced development. As such, R&D constitutes a major part of the entire defense weapons acquisitions process, called "Program 6," in the DOD Future Year Defense Program.[37] Congress authorizes and appropriates funding for research, development, test and evaluation under the auspices of Program 6 in the annual defense budget.

DOD has broken up Program 6 into five distinct categories, shown in Table 1, roughly corresponding to the research definitions given earlier in this chapter.[38]

**Table 2. *Program 6 Development Phases***

| *PE* | *Category* | *Distinguishing feature* |
|---|---|---|
| 6.1 | Research | Basic research |
| 6.2 | Exploratory Development | Applied research for feasibility |
| 6.3 | Advanced Development | Develops components for system tests |
| 6.4 | Engineering Development | Contract definition |
| 6.5 | Management and Support | Operational development |

It is tempting to view the development cycle of weapon systems as a simplistic, well-ordered process such as shown in table 2. And that is what is frequently done by those who have not performed laboratory research, for this is the way that scientists and engineers have communicated the way they have performed creative research. However, in reality the lines between the different areas are blurred and may overlap—or there even may be gaps in the knowledge, where experimental results may show a promising phenomenon, but no explanation is yet known. Or the process may involve parallel (concurrent) developments that transcend one or more of the Program 6 Development Phases.

The quintessential example of this is the "bunker buster" kinetic energy weapons fielded during *Desert Storm*. The idea of accelerating a massive body to burrow deep into the ground to destroy buried command bunkers was not discovered during a 6.1-funded project; rather, it was borne years earlier from such concepts as "fire from heaven"—massive metal rods gravitationally accelerated from orbital heights to be used for precision strike. The "bunker buster" idea went directly to a 6.3 "Advanced Development" stage to prove the concept. And within *days* the weapons were hastily assembled and shipped off to combat.

The point is that the development phases in table 1 reflect an established *process*, set up and validated by DOD. The largest percentage of money may be spent during the last three phases (over 80 percent), but the *tone* of the program—the direction and promise of a new capability—is set during the first stage (6.1), where until 1966 was typically the first 5 percent of the money spent.[39] Table 3 shows the estimated FY96 Research, Development, Test and Evaluation (RDT&E) authorizations, with research (6.1) comprising 3.5 percent of the total RDT&E in FY96, *a decrease of 30 percent from the mid-1960s*.[40]

As the initial stage of development is the most remote (both conceptually and practically) from the warfighter (the person who either directly or indirectly controls the funding), it is absolutely imperative to lay a firm 6.1 basic research foundation. Otherwise, the warfighter—the person who demands results today, but yet is faced with funding investments that may not pay off for years—will always choose to forego long-term research for quick, short-term results.

Table 3. *RDT&E Appropriations (billions)*

| *Development Phase* | *FY94* | *FY95* | *FY96 (est)* |
|---|---|---|---|
| Research | $1.167 | $1.227 | $1.214 |
| Exploratory Development | $2.691 | $3.070 | $2.816 |
| Advanced Development | $6.208 | $4.339 | $3.796 |
| Demonstration and Validation | $2.697 | $4.325 | $4.229 |
| Engineering & Manufacturing Development | $7.334 | $8.930 | $8.759 |
| RDT&E Management Support | $3.369 | $3.436 | $3.305 |
| Operational Systems Development | $11.242 | $10.188 | $10.213 |

Therefore, transistioning from 6.1 to 6.2 research —basic research to exploratory development, the abstract to the possible—is crucial for ensuring that technology continues to grow.

Note from table 3 that the lower percentage of 6.1 research (3.5 percent) to the total RDT&E in FY96 as compared to the mid-1960s (5 percent) shows *quantitatively how long-range, basic research is regarded as a lower priority by decision makers—in fact thirty percent lower than in the past.* The percentage of exploratory research (6.2) has also dropped, from 12 percent in 1966 to 8.2 percent in 1996.[41] Over a span of 30 years, the total S&T (6.1 and 6.2) percentage has dropped 31 percent relative to the rest of RDT&E. For the decision makers, this vividly "puts their money where their mouth is."

To put this is perspective, it is useful to remember that the R&D investments made in the 1960s produced the technology that won both the Cold War and the Gulf War. There is much to be said for not having a "hollow force," and being able to purchase the weapon *du jour*; but what does this astonishing decrease in S&T percentages say about how our decision makers view tomorrow? Or their vision?

## THE ROLE OF LONG-RANGE RESEARCH

It has been said that art mimics life. In movies and literature, scientists are never portrayed as benign. They are shown as either purveyors of doom or saviors of the world, whipping up a solution to the hardest problem. In the real world, science is often viewed with the same naiveté.

Such is the stigma associated with the role of science to warfare. It is the advancements in chemistry which made possible explosives and various chemical gases, the horrible weapons of mass destruction; rudimentary aeronautical engineering brought advances in airplanes, which were used first as observation platforms, then to drop bombs; the physics brought the world nuclear fission, which paved the way for first the atomic bomb, and then as a trigger for the world's most destructive weapon, the hydrogen, or fusion bomb.

But the same science which brought destruction also served to counter the march of ever-increasingly sophisticated weapons fielded by our adversaries. For example, anti-missile technology drew on advances in mathematical algorithms, material properties, structural designs, energetic fuels, control theory, sensor technology, and a host of other areas. The antimissile missile offered a chance to counter the missile. Each new advance seemed to bring not only the possibility of a new weapon, but the opportunity to negate, or at the very least render neutral, an older weapon.

## EVOLUTIONARY VERSUS REVOLUTIONARY ADVANCES

Some have chronicled major advances in science and have called them "revolutionary." It is claimed that the world has changed in many ways, and that these advances in science are so fundamental to the treatment of warfare that they have even caused the very nature of war itself to change. Thus grew the concept of "Revolution in Military Affairs," or RMA. Each RMA is associated with the discovery and subsequent application of a groundbreaking scientific or technological advance. For example, the bow and arrow was a vast improvement over the spear; the rifle over the bow and arrow; the introduction of tank warfare; the airplane; the atomic bomb; and some say today the computer.

But the world, the nature of warfare, and in particular the essence of science itself does not instantly cause change.[42] It is not the revolution in science and technology that causes a revolution in military affairs, it is the *evolution* of science and technology. Immediate, world shaking events that profoundly change war are *not* invented by a team of brilliant scientists and engineers, whipped up on the spot to handily defeat the enemy. Their weapon served only to accomplish one goal. The *real* change occurred over a much longer term, but if this is true, then what about the atomic bomb?

The Manhattan atomic bomb project at Los Alamos during World War II was based on work performed by physicists who discovered the process of nuclear fission. The science and technology that resulted in the atomic bomb took only 3 years to come to fruition once the Manhattan project was established—an astonishingly short time, especially considering the massive technological infrastructure that not only had to be assembled, but also had to be *invented*, concurrently, along with the breathtaking advances in basic science that had to be made. The Manhattan project by any measure was an astounding success.

This work resulted in two different nuclear devices, one based on the implosion or compression method and the other based on a "gun" or impact method. But the true Revolution in Military Affairs was not the invention of these weapons, or even the fact that one device of each kind was used on Japan to expedite the end of the war. It was the *evolution* of the atomic bomb, and its subsequent diffusion through first the American, then the Soviet Union military infrastructure, that made this a true RMA. Nuclear weapons had to rely on the huge buildup of bomber aircraft and the invention of ballistic missile technology in order to profoundly change the world. Advances in flight control, stability, propellants, triggers and a myriad of interdisciplinary fields were brought together and optimized. It was the growth and maturation of the nuclear weapons infrastructure—including the delivery systems, the warhead design, and production facilities—that caused the Revolution in Military Affairs.

In the same manner, it wasn't the invention of the bullet that overcame the bow and arrow; it was the application of mass production techniques to effectively produce the guns that caused the RMA.

Another example is the addition of precision-guided bombs to aircraft. It wasn't the laser, or the homing device, or even the infrared tracker that precipitated this high-leverage, "force multiplying" invention. It was a synthesis of technologies and discoveries, all matured over a period of time. A period of time where *long-range investments* were allowed to pay off.

The point to this whole discussion is that scientific discoveries alone cannot profoundly affect warfare. Recall the "invention to innovation" time-scale in an earlier section—it took 27 *years* to fully develop the zipper. So it is the *evolution* of these discoveries, matured by interbreeding advances in other fields, that resulted in an innovative use.

*And that is the strength and role for long range research in DOD—to first produce those new discoveries, and then to feed them into the technology base. For this enables the discoveries of today to become the practical weapons of tomorrow.*

## TRANSITIONING LONG-RANGE RESEARCH

In their *New World Vistas* report detailing critical areas targeted for investment, the Air Force Scientific Advisory Board posited seven general guidelines.[43] These guidelines were meant to put the revolutionary concept in context for warfighters who expect miracles from long range research:

- **The relationship between revolutionary and evolutionary concepts is complex and complementary.** As shown above, rarely is a scientific discovery completely revolutionary, in the sense of causing a paradigm shift—that is, altering the world-view, or focus of how "business" is conducted. Instead, it is the associated advancements in technology, the nonlinear cross-feeding of inter-related fields that allow for the evolutionary growth that may someday be perceived as revolutionary. For what is revolutionary to one may be viewed as evolutionary to another—one may merely have not been able to adjust to the changes.
- **Revolutionary ideas often point the way to later applications which are far more useful than the original idea.** Forecasting the eventual use of a technology is beyond the ability of anyone because of the myriad casual forces and the inability to know fully all the boundary conditions necessary to make an accurate judgment. This philosophy was the foundation for the Bush administration's policy of *not* implementing a national industrial policy. It was felt that the government could not pick the "winners" in a free market

environment, because the commercial marketplace would (and should) determine both the winners and the losers.

The standard example is GPS—Global Positioning Satellites. No one accurately projected the myriad uses of this now critical resource. Another, but more infamous anecdote of the inability of individuals to predict the application of a technology is the "aerogel" case at the Lawrence Livermore National Laboratory (LLNL). As one of the two nuclear weapon design laboratories run for the Department of Energy, LLNL developed "aerogel" as part of its nuclear weapon program. A wispy, transparent substance made partially from seaweed, aerogel is extremely light yet very strong. Efforts at LLNL to transition aerogel to the commercial sector failed because no one could think of a viable commercial use of the substance. However, when turned over to the commercial market place, a heretofore unthought of use appeared—using the aerogel in cigarette lighters to make ultralight but extremely strong reservoir for lighter fluid.

- **Early application of revolutionary concepts should not be required to be complete and final weapon systems.** Expecting too much too soon from an unproven concept may lead to disaster. Scientific experiments are littered with stories of failure, and applying requirements to revolutionary concepts may actually do more harm than good.

The SSTO (Single-Stage-To-Orbit) spacecraft is an excellent example of what could happen if operational requirements are levied on a revolutionary concept. The concept of building a fully reusable spacecraft that would only have one stage is driven not by magic, but by the rocket equation, a sophisticated expression of momentum conservation. There are several formidable obstacles to building an SSTO (mass fraction, heat resistant materials and propulsion among them), but the fact remains that although the engineering problems are huge, they are not impossible. Scientifically there is no reason for an SSTO not to work; *technically* there are plenty of reasons.

The point is that for such a formidable project, the only rationale way to solve the problem is to first show that an SSTO can work—that is, that the problem is technically possible. Any attempt to make the SSTO a weapon system without first showing that it works dooms the project to failure. The requirements levied on a weapon system are different from the easier problem of showing it works. This is precisely what happened to the National AeroSpace Plane (NASP), the "Orient Express"—an SSTO capable of carrying passengers from one part of the world to another, as envisioned by President Reagan. By the time the NASP became a program, it was overburdened with military requirements in order to gain the support across the military services, and the requirements demanded enormous capability from a plane when it was not even known if it was technically possible to get to orbit.

And that is why the new NASA SSTO project is so exciting—it is an attempt to prove a concept, build an "X-rocket" much like the X-series of experimental aircraft built in the 1950s and early 1960s *to prove the technology*. No military requirements were levied on the X-planes; it was only after the technology was proven possible that the requirements came.

A modern example of how this succeeds is with the F-117 stealth fighter. Originally built as a technology demonstrator under the classified *Have Blue* program, it was only after the feasibility of stealth was proven that the F-117A was built.

Just as there are exceptions to every rule, there is an exception to this one as well. Certainly, if the very existence of the nation is at question, such as in World War II, when it was thought that Germany was close to obtaining the atomic bomb, then a last-resort crash program is justified. Even in the post-Cold War era, there may be some threats that are of the magnitude that requires disregarding this axiom—such as chemical or biological attacks by a rogue state.

- **Identification and development of revolutionary concepts require intuition, innovation, and acceptance of substantial risk.** The aircraft industry has a rich history of funding high risk development efforts for experimental aircraft.[44] The industry realizes that to be competitive, it must take risks to advance the state-of-the-art; evolutionary advances of improvements may strengthen the profitability of a product line and is commendable in a "quality" environment, but unless new ground is broken, a competitor may change the state of the industry overnight. The Boeing Company undertook substantial risk that involved immense technological challenges to develop the 777 aircraft, but the upturn of the commercial market coupled with increasing demand for the new aircraft validated their gamble of accepting this risk—a gamble involving nearly 5 billion dollars for Boeing.[45]

The antithesis of this is the rocket launch industry, primarily because of a hedge "to protect any investment against losses."[46] Often viewed as "risk adverse," some in the launch industry argue that the cost of losing a launch vehicle (on the order of tens or even hundreds of millions of dollars), and also losing the associated payload (as much as a billion dollars) prohibits risk. Therefore, to decrease the risk, more money is spent on the launch system to increase the *reliability*, but this requirement in turn makes the launch more *risky* because of the increased amount of money that now at stake if the rocket is lost. So even *more reliability* is needed to decrease the risk, which paradoxically *increases* the risk.

And one can see easily see, this tightly coupled, ever-increasing death spiral of risk begetting reliability in the launch industry only results in more cost for the consumer—in this case, the U.S. Government and the commercial marketplace who desire to exploit space. Risks *must* be taken if any substantial advance is to be made. This is true of every other industry, and there is no reason why this should also not be true for the space launch industry—and all of science and technology research—as well.

- **We must be prepared for a failure rate greater than 50 percent.** During the early development years of the U.S. intercontinental Ballistic Missile (ICBM) program, rockets routinely failed at a rate approaching 60 percent.[47] In the scientific field, scientists routinely learn more from their failures than from their successes. With the understanding that defense-driven science and technology may have literally millions of dollars funding an advanced technology demonstration, a policy of allowing for no failures will only result in costly, meaningless "performances" that do little to advance the true state-of-the-art. Emphasis will be put on reducing failure and not "pushing the envelope," resulting in slow, evolutionary increases in performance, rather than in true revolutionary advances.
- **Most revolutionary ideas will be opposed by the majority of decisionmakers.** Many decisionmakers have risen to their position because of competence in some area, be it managerial skills, technical savvy, or the ability to win a political appointment. When they are confronted with a new situation it is easy to fall back on what has gotten them through past situations. In Boleman and Deal's text, *Reframing Organizations*, they elegantly states, "Too often [leaders and managers] bring too few ideas to the challenges that they face. They live in psychic prisons because they cannot look at old problems in a new light and attack old challenges with different and more powerful tools—they cannot *reframe* . . . the lack of imagination is a major shortfall between the reach and grasp of so many organizations."[48] Another way of expressing this is

that "you lead what you know," and therefore new ideas, especially revolutionary ones, are viewed as a threat to the status quo.[49]

- **We must remember that science and science fiction are related only superficially.** Here the Science Advisory Board tries to ensure that the defense official who sees a flashy pseudo-science feat achieved in the movies (or for that matter reads about it in a book) does not automatically try to develop that capability for DOD. Their point is that it is the realm of science fiction to break the laws of physics, not the acquisition official.

However, Dr. Gene McCall, the chairman of the AF Science Advisory Board, has pointed out that there *are* some things that can be gleaned from science fiction. He calls this the "possible" and the "impossible" aspects of *Star Trek*: some "possible" uses of technology include the communicator (that can be used as a point-to-point communication device anywhere in the world), the medical "tri-corder" (that can give internal medical assessments of humans without touching them), and the universal translator. Certainly these examples of technology do not violate any of the known laws of science and might be possible to manufacture given sufficient technological advancement.[50]

In the same vein, however, the "impossible" aspects include "warp drive" (traveling faster than light), the use of a "transporter" (beaming matter, including people, instantly from one spatial point to another), "sub-space communication" (faster than light communication), and "time travel." Certainly if a fundamental discovery was made that overturned the laws of physics, these aspects might be revisited, but the chances of exploiting them for practical use is vanishingly small.[51] However, it should be noted that many scientists and engineers were steered into their profession because of the lure of the fantastic, be it possible to achieve their dreams or not.

## SUMMARY

The Department of Defense has a tradition of support for long-term research. A majority of the defense establishment may only grudgingly acknowledge the critical role that research plays, but the importance of scientific advances is seen throughout the modern arsenal: stealth platforms, high precision weapons, extremely accurate navigation, communications—and the list goes on.

The weapons of today are grounded in basic research performed 30 years ago—at 5 percent of the RDT&E budget in the 1960s. Today, that percentage has fallen by a third. The question that policy makers will have to face is did they do enough today for tomorrow?

### NOTES

1. National Science Foundation, Federal Funds for Research and Development and Other Scientific Activities, vol. XV, (Washington, DC: U.S. Government Printing Office, 1966), 60.

2. Richard H. Stephens and Eston T. White, eds., *Science and Technology* (Washington, D.C.: National Defense University, 1983), 12.

3. National Science Foundation, "Annual Survey of Industrial Research and Development," 1995.

4. Director of Defense Research and Engineering, *Defense Science and Technology Strategy* (Washington, DC: Department of Defense, September 1994), 15.

5. White House Committee for National Security, National Science and Technology Council, *National Security Science and Technology Strategy* (Washington, DC: Office of Science and technology Policy, Executive Office of the President, 1995), 7.

6. Stephens and White, 12.

7. White House Committee for National Security, 7.

8. This absolutely drives the accounting people bonkers. Scientists cannot point at a research project and say "The project is basic research at this point and over here it has tranisitioned to exploratory development." The process is

a continuum, and is not discrete.

9. William L. Miller, "A Broader Mission for R&D," *Research Technology Management*, November-December 1995, 40.

10. Nathan Rosenberg and L.E. Birdzell, *How the West Grew Rich* (New York: Basic Books/Harper Collins, 1986), 244.

11. Ibid.

12. Actually, some advances even in the scientific field still happen by accident. For example, the discovery of the background microwave radiation in space was totally unexpected. And there is even truth to the stories that some scientists still conduct their research as an "art:" Look to the advances in room temperature superconductivity, where no rigorous theory yet exists, and many of the advances are still made by insight.

13. Rosenberg and Birdzell, 243.

14. Isaac Asimov, *Asimov's Chronology of Science and Discovery* (New York: Harper & Row, 1989), 338-371.

15. Donald L. Losman and Shu-Jan Liang, *The Promise of American Industry: An Alternate Assessment of Problems and Prospects* (Westport, CT: Quorum Books, 1990), 104.

16. Rosenberg and Birdzell, 254.

17. Daniel Boorstin, *The Americans: The Democratic Experience* (New York: Vintage Books, 1974), 538-539.

18. Nathan Rosenberg and L.E. Birdzell, *How the West Grew Rich* (New York: Basic Books/Harper Collins, 1986), 247.

19. Hendrick Smith, *Re-Thinking America* (Random House, NY, NY: 1995), 67-71.

20. Genevieve J. Knezo, "Defense Basic Research Priorities: Funding and Policy Issues" (Washington, DC: Congressional Research Institute, 90-506 SPR, October 24, 1990), 12.

21. Executive Office of the President, Office of Science and Technology Policy, *National Security Science and Technology Strategy* (Washington, DC: U.S. Government Printing Office, 1995), 64-65.

22. Ralph Sanders, *Defense Research and Development* (Washington, DC: Industrial College of the Armed Forces, 1968), 73.

23. Ibid.

24. Ibid., 33.

25. Ibid., 34.

26. Ibid., 34.

27. Herbert N. Foerstel, *Secret Science: Federal Control of American Science and Technology* (Westport, CT: Praeger, 1993), 2.

28. Ibid.

29. Paul Kennedy, in *Preparing for the Twenty-First Century* (Random House, NY, NY: 1993), 293, argues that although military funding may have diverted some money from the civilian sector, many technological advances were made because of military R&D that might not have been made otherwise. This is a "boot strap" argument, showing how military R&D spin-offs have helped society as a whole. Another way to look at this is to consider what the money would have been used for if it had *not* been used for military R&D. With the waste, fraud, and abuse present in myriad non-DOD programs, it is highly likely the money would have been squandered

30. Sanders, 36.

31. Ibid., 36.

32. This situation is inherent of almost all review committees that exist today in Washington, DC. The typical process of selecting committee members is to pick a large majority of *past* committee members—those that have served on similar committees in the past. This results in self-perpetuating views—the members of committees drawing upon recommendations that they made in the past. Another common tactic is for committee members to be picked because of their preconceived notions, especially if it happens to coincide with the views or agendas held by those tasked to create the "unbiased" report.

33. Sanders, 36.

34. Ibid., 37.

35. Ibid., 73.

36. Office of Management and Budget, Executive Office of the President, *The Budget of the United States Government* (Washington, DC: U.S. Government Printing Office, 1995).

37. Sanders, 10.

38. Ibid.,18.

39. Ibid.,18.

40. FY96 RDT&E: Aerospace Fact Handbook, 63.

41. Sanders, 18.

42. Air Force Scientific Advisory Board, *New World Vistas: Air and Space Power for the 21st Century*, report to the Secretary of the Air Force and Air Force Chief of Staff (Washington, DC: USGPO, 1995), 13.

43. Ibid., 14.

44. Ben Rich and Leo Janos, *Skunk Works: A Personal Memoir of My Years at Lockheed* (Little, Brown and Company, New York: 1994), 112.

45. Laura D'Andrea Tyson, *Who's Bashing Whom? Trade Conflict in High-Technology Industries* (Washington, DC: Institute for International Economics, 1992), 163.

46. U.S. Congress, Office of technology Assessment, *The National Space transportation Policy: Issues for Congress*, OTA-ISS-620 (Washington, DC: U.S. Government Printing Office, May 1995), 15.

47. Dr. Michael D. Griffin, General Manager, Advanced Systems Group and Executive Vice President, Orbital Sciences Corporation, in remarks before the National Security Industry Study symposium on "Space: Getting out of the Box," held at the Crystal City Gateway Sheraton, Arlington, VA, on March 6, 1996.

48. Lee G. Bolman and Terrence E. Deal, *Reframing Organizations* (San Francisco: Jossey-Bass, 1991), 4.

49. Remarkably, this propensity of "putting up shields" to fend off revolutionary ideas is just as likely to be abused by self-claimed "forward thinkers" as anyone else. One reason is that perhaps the self-called forward thinkers have realized they are not as competent in their profession as others, and this is a way to compensate for their shortcoming. And when faced with a truly revolutionary idea, they cannot become dispassionate.

50. To physicists this is known as "just an engineering problem."

51. The common argument for championing the "impossible" technologies often found in science fiction series such as STAR TREK is that the same skepticism was brought up about landing a man on the moon. The difference of course is that landing a man on the moon violated no physical laws—only "common sense" at the time.

# 4. THE PRESENT DEFENSE RESEARCH STRATEGY

> We are not the only nation with competence in defense science and technology. To sustain the lead which brought us victory during *Desert Storm* . . . recognizing that over time other nations will develop comparable capabilities, we must . . . invest in the next generation of defense technologies.
>
> William J. Perry

## THE FOUNDATION OF DEFENSE SCIENCE AND TECHNOLOGY

Military science and technology (S&T) must contribute to the national security goals of the United States by being responsive to new threats, challenges, demands and opportunities.[1] To ascertain specifically how these national goals are addressed, the Joint Staff was tasked by DOD to set up a process to identify future joint warfighting capabilities. The capabilities are evolved from the Joint Requirements Oversight Council (JROC), an executive-level council made up of the vice chiefs of staffs of the military services. The JROC capabilities serve to guide the S&T investment decisions made by the defense research establishment, specifically by the Director of Defense Research and Engineering (DDR&E). This highly interactive process strongly couples future military capabilities with national policy goals.

The JROC has identified five future joint warfighting capabilities, given below.[2] These capabilities lay the foundation for the U.S. National Security Science and Technology Strategy, which then sets the requirements that drives the S&T contributions for military needs. Once the military needs are established, a science and technology strategy may then be formed to satisfy these needs.

It is satisfying this S&T strategy through the coupling of long-range and short-range research that is the subject of this chapter. Specifically, problems with the strategy in transitioning from long-range to short-range research will be addressed—and that ultimately affects how the warfighter will have access to the latest advancements in science and technology. But before this coupling may be explored, it is absolutely necessary to know the rationale underpinning defense S&T. The following five future joint warfighting capabilities have been determined by the JROC process.

## FUTURE JOINT-WARFIGHTING CAPABILITIES

**To maintain near perfect, real-time knowledge of the enemy and communicate that to all forces in near-real time.** Operational forces must know precisely the enemy's location and capability. This information must be transmitted to warfighters in a near-real time manner, along with information as to the status of their own forces and the spectrum of possible responses to defeat the enemy. Included in this information are dynamic data about the weather, geography, topography, the political situation, and the location of civilians.[3]

The delivery systems for these data—battlefield surveillance, information management, and communications—require a highly coupled, optimized, and efficient means of transmission.[4] If these data are not transmitted in a near real-time manner, then disastrous consequences may occur. The shoot-down of Captain Scott O'Grady's F-16 over Bosnia has recently been acknowledged as stemming from a failure of providing this type of information in a timely manner—information had been obtained of the presence of antiaircraft radar and missiles but had not been transmitted to the operational warfighters.

**To engage regional forces promptly in decisive combat on a global basis.** Regional forces depend on two axioms to bring them rapidly to victory: mobility and decisive combat. Mobility includes rapidly introducing both the operational forces and their supplies into combat.[5]

To increase mobility, advances in increasing the lift capability of air, land, and sea vehicles are performed in concert with decreasing the weight of what is carried. Science and technology can create advances in propulsion, avionics, materials and structures (for reducing weight and increasing strength), and fuels. Computer simulations can add to optimizing the delivery systems and their routes, as well as numerically testing innovative approaches to solving these dynamically changing optimization problems.[6]

Decisive victory in combat may be enhanced by feeding the warfighter near real-time information about the battlefield environment, including the location and capability of enemy forces. This includes rapid identification and tracking of targets, as well as knowledge and location of friendly forces. It is enhanced by providing the warfighter with a precision strike capability from an extended range, fielding anti-weapon armor, and giving the warfighter an ability to neutralize or destroy the enemy's aircraft and missiles.[7]

There are other means to feed the warfighter timely information. New technologies, such as the introduction of unmanned surveillance vehicles, and by deploying sophisticated sensors into the air, land and sea, can provide a means of obtaining intelligence without putting the warfighter in harm's way.[8] Science and technology also provide the means to rapidly respond to new, unique threats—such as deeply buried command and communication bunkers in Iraq, numerous tunnels in North Korea, or the millions of land mines in Bosnia.

**To employ a range of capabilities more suitable to actions at the lower end of the full range of military operations which allow achievement of military objectives with minimum causalities and collateral damage.** From 1950 until the early 1990s the U.S. military was built and positioned to respond to an overwhelming monolithic threat—the expansion of communism fueled by the Soviet Union's military machine. With the fall of the Soviet Union and the subsequent disappearance of an overpowering foe, the main threat forecast to the United States over the next several decades will come from small regional conflicts and the dangers associated with conducting Special Operations, assisting humanitarian efforts, and conducting peacekeeping operations.[9]

Achieving military objectives while conducting special operations or while providing humanitarian assistance calls for using another metric for measuring success—obtaining minimum causalities with little or no collateral damage. Nonlethal weapons such as high-power microwaves, "sticky foam," or sonic weapons, among others, show great promise for achieving military objectives. The use of nonlethal weapons that have the capability to neutralize, stun, disable, disorient, or confuse the enemy without having lasting effects is therefore highly desirable.[10]

However, intrinsic to these "low end" military options is the real threat to the U.S. warfighter. These hostile situations can be dangerous, as crowds may turn against them, or they may open themselves up to attacks by terrorists. In addition, exotic weapons such as chemical and biological warfare may be used against U.S. forces. Advances in information technology can be used to give the warfighter a heightened degree of battlefield awareness; biological and chemical advances may afford protection that heretofore had been impossible to obtain.[11]

**To control the use of space.** The United States holds a monopoly in the free world's (i.e., excluding Russia) use of space, having launched 1,018 spacecraft to Earth orbit or beyond from 1957 to 1994, compared to 175 launches for the European Space Agency, Japan, China, India, Israel, France, Italy, Australia, and the United Kingdom *combined.*[12] However, this is less than *half* the number that the former Soviet Union had launched in the same time period. Furthermore, the U.S. continues to launch approximately half the number of spacecraft Russia launches today.[13]

However, no matter what the present relative U.S. ranking, in *Desert Storm* the importance of space to national security was brought solidly to the front. Space assets provided the United States with intelligence, positioning information and communications that greatly added to its ability to win the war. The United States clearly held the "high ground," and if the Iraqis had had access to the same intelligence as the United States, the war might have turned out differently.

In fact, this almost happened. The French, while participating as part of the Coalition Forces in the Gulf War, provided Spot satellite overhead data (10-meter resolution) to the Iraqis. With great credit to U.S. prodding, the French "reconsidered" what it was doing and cut off Iraq's access to a potentially dangerous windfall of intelligence data. This demonstrates just how fragile the U.S. claim to the "high ground" of space actually was.

Currently, numerous countries have launch capabilities and have operational satellites in place. The European Space Agency *Ariane*, the Russian *Proton*, the Japanese *H-2*, the Chinese *Long March*, and other countries' launch vehicles (such as India) are all capable—or will be in the near future—of launching satellites to low earth orbit or higher. Countries as disparate as Indonesian have communication satellites to orbit. With the proliferation of smaller satellites weighing mere hundreds of pounds—or even *microsatellites* weighing *tens* of pounds—and the availability of "low cost" launchers (in the $10-million range), more countries will be able to launch and control their own satellite assets.[14] And even if a country does not have its own indigenous space capability, U.S. commercial companies, under a recent Presidential Decision Directive, are now allowed to sell hyperspectral satellite data to foreign countries down to one meter in resolution.

With these emerging capabilities, the United States must position itself with the judicious use of science and technology to rapidly and efficiently control the use of space during future conflicts. This includes protecting U.S. military satellites and launch capabilities, denying foreign countries the ability to threaten these assets, and even protecting friendly non-military satellites, such as communication links. This may be accomplished by investing in technologies that optimize launch (either through low-cost expendable launchers or developing a single stage to orbit capability), substantially reducing the time from satellite design to launch (from 10 years to 18 months) and developing an antisatellite capability.[15]

**To counter the threat of weapons of mass destruction and future ballistic and cruise missiles to the CONUS and deployed forces.** The United States lived through the threat of global annihilation, brought on by the possible use of nuclear weapons during the Cold War. Enormous quantities of warheads existed on both sides. Unclassified sources claimed that as late as 1993 the numbers ranged from 10,000 weapons in the U.S. inventory to over 30,000 in the four major countries that made up the former Soviet Union.[16] But the terror invoked by the use of nuclear weapons actually worked successfully to end the Cold War: no atomic bombs were dropped and the two superpowers did not go to a "shooting" war, at least not directly against each other. So one may conclude that the very existence of nuclear weapons brought about stability.

However, with end of the Cold War and the slow dismantlement of the superpower's nuclear arsenal, there still exists a very real danger of nuclear weapons and other weapons of mass destruction. In fact, there now may be even more of a danger, in that the "stability" of a bi-polar world has evolved into that of a single superpower surrounded by allies, regional wackos, and an undercurrent of chaotic factions existing in third world countries.

Nuclear weapons are not difficult to make, in a relative sense.[17] The basic physics behind nuclear weapons has been known for several decades: slap two pieces of fissionable material together to make a "critical" mass of fissionable matter, which then releases its energy in an atomic explosion. A hydrogen (fusion) bomb, up to a thousand times more powerful than an atomic (fission) bomb, works simply by rapidly fusing hydrogen together to make helium. The fusion process is actually *driven* by first detonating an atomic bomb, but the fact is that producing a hydrogen bomb is a well-known, relatively simple problem in physics.[18]

The heart of the matter in building a nuclear weapon lies in the engineering details—the materials and correct manufacturing processes that lead to a completed device. This demands a sophisticated infrastructure. In practice a nuclear weapon is extremely difficult to build, but it is quite within reach of several countries—such as Iraq or North Korea, for example.

And chemical and biological weapons of mass destruction are easier to manufacture than nuclear weapons. Although certain growth media are need to produce biological weapons, chemical and biological weapons do not require the sophisticated supporting infrastructure as does their nuclear counterpart. Witness the recent gas attacks on the subway system in Japan.

Therefore, it is safe to assume that weapons of mass destruction will not disappear, and that the United States must be prepared to both stop their proliferation and to be able to actively deny their use. Science and technology may play a role in this, through sophisticated detection mechanisms such as neutron activation of materials, or in neutralizing their effects, such as through the focused application of concentrated ozone, or intense X-rays to kill biological entities.[19]

### THE CHARGE TO DOD

With these five future warfighting capabilities identified, the President has charged the U.S. Department of Defense with ensuring our future national security remains intact:

> The Administration is committed to a sustained investment in the technology base needed to ensure that our nation maintains the best-trained and best-equipped forces in the world. Our investment strategy involves long-term research as well as near-term applications as it is only in hindsight that we are able to discern the revolutionary military capabilities provided by breakthroughs such as radar, digital computers, semiconductor electronics, lasers, fiber optics, and navigational systems capable of great accuracy.[20]

As the executing agent for the DOD science and technology program, the Director of Defense Research and Engineering (DDR&E) uses the five future capabilities to focus the overall content, quality, and direction of defense-related R&D to accomplish the administration's goals. The DOD S&T program is organized in three overlapping categories: basic research, exploratory development (applied research), and advanced technology development. The quintessential core of military strength is anchored on this foundation. This remainder of chapter will examine the present defense research infrastructure and illuminate its strengths and weaknesses.

## NATIONAL SECURITY SCIENCE AND TECHNOLOGY STRATEGY

### S&T CONTRIBUTIONS TO MILITARY CAPABILITY NEEDS

The success of the U.S. National Security Strategy is directly linked to the nation's commitment to science and technology.[21] Specifically, the success ultimately depends on the efficiency and timely exploitation of S&T to develop and deploy sophisticated weapons systems to counter and defeat the enemy. Some authors have pointed out that this policy involves the United States embracing a "technology-intensive" strategy rather than a "manpower-intensive" strategy.[22] The differences may seem slight, the but consequences are large.

As an example of a technology-intensive strategy, the United States brilliantly responded to the Soviet ICBM threat of immense boosters capable of lifting multi-megatons to fractional (ballistic) orbits. The United States developed highly accurate warheads with relatively small nuclear yields, on the order of hundreds of kilotons. It was a classic response of using finesse to counter brute strength: what the United States lacked in brute strength (massive rockets and megaton-class warheads) it overcame with lower-yield weapons lifted by modest size rocket boosters.

But the U.S. reliance on technological finesse went further than merely responding with sophisticated weaponry. A philosophy of "quality over quantity" was constantly espoused and adhered to. It became a national mantra and permeated our culture. The U.S. response to the massive Soviet in Western Europe was based on this philosophy. National policy dictated that the United States would respond to a "manpower intensive" threat by investing in our educational system, increasing the funding of our scientific infrastructure, and strengthening our industrial base.[23] This philosophy explicitly surfaced through many programs: the Apollo program and the Strategic Defense Initiative, to name a few.

Thus was defined the US's "technology-intensive" paradigm for responding to "manpower-intensive" threats. In addition to its ethereal benefits of precisely countering the Soviet problem, it also allowed policy makers to avoid the messy issues associated with wide-spread conscription.[24] The timely and intelligent use of technology was viewed as a move away from the "attrition based" warfare of the past to "information based" warfare of the future—exploiting technology to produce unmanned, precision, brilliant weapons. This is stated eloquently by Gansler, "Constant development and deployment of new weapons will be needed to stay ahead."[25]

Therefore, a principal challenge of the United States maintaining its military advantage is to ensure technological superiority over present and future adversaries.[26] In maintaining this superiority, a major requirement is that defense-related research must ultimately satisfy the national vision.

A "bottom up" structure for achieving this requirement is to ensure that basic research advances science and technology. Basic research must be "focused" but done so in a general enough field that it is likely to produce advances in the S&T desired for future weapons systems. For example, it is extremely unlikely that performing basic research on economic models will result in a significant advance in materials science for submarine propellers; or it is unlikely that performing basic research on fresh cow milk will produce an advance in rocket propellants. However, the *techniques* pioneered in these disparate fields may well be applicable to defense-related S&T. For example, techniques solving a diffusion equation used for stock market analysis may be applicable to the flow of laser energetics. Therefore, advances in S&T may contribute to advanced technology demonstrations that exploit new or refined discoveries.

Once a successful S&T demonstration has validated a new weapon concept, the system undergoes development to focus and optimize the system. If more than a few of these systems are required in the field, then the system undergoes a production phase so that it may be transitioned to the field to exploit its operational capabilities.[27] The operational capabilities of this new weapon system should satisfy the national security S&T strategy, if in turn it is to ultimately satisfy the vision set out by national policy makers.

## VISION

The Department of Defense Science and Technology Vision has been established to provide technological superiority for our nation's military services. Encompassing all defense-related research conducted by DOD, the Director of Defense Research and Engineering published the following vision in it's September, 1994 *Defense Science and Technology Strategy*:

> Develop and transition superior technology to enable affordable, decisive military capability and to enhance economic security.

This vision recognizes that maintaining technological superiority is not enough; it is a necessary but not a sufficient condition to ensure our military superiority. Coupled with the necessity for ensuring technological superiority is the dual condition of producing *affordable weaponry* and *enhanced economic security.* These two conditions extend the vision to account for the post-Cold war environment, demanding that DOD pursue research with a new mindset. Further, the vision recognizes that both *economies of scale* and *technological innovation of the commercial sector* play a vital, central role in defense-related research.[28]

The vision's bottom line is to improve productivity and reduce costs while producing effective weapon system's capabilities.

## SCIENCE AND TECHNOLOGY STRATEGY

The U.S. National Security Science and Technology Strategy stands on three pillars. These pillars define the foundation of our national power, and lie at the heart of the Clinton administration's definition of national security:[29]

- The readiness and capabilities of our military forces
- Our engagement with other nations to prevent conflict from occurring
- The strength of economy

These pillars are maintained by focusing our investments to allow for their growth and health. The investments that the nation can focus on are neither monetary nor personnel-driven; rather, they are in specific, critical *high-leverage areas* that can ensure that the nation continues to be strong. The Director of Defense Research and Engineering has identified four critical investment areas:

- Maintain technological superiority in warfighting equipment
- Provide technical solutions to achieve future joint warfighting capabilities
- Balance basic research and applied technology in pursuing technological advances
- Incorporate affordability as a design parameter.

Investment in each of these areas, discussed in detail below, are critical for ensuring the health of the national security science and technology strategy.

**Maintain technological superiority in warfighting equipment.** The United States cannot afford to fight its next war with the technology from the last war. The rate of technological growth is accelerating throughout the world—as measured by R&D expenditures (including universities, industry and government), after adjusting for inflation, the total outlays for R&D more than doubled in the United States alone between 1960 and 1984.[30]

Nations can now obtain technology from U.S. and other commercial markets that only a few years ago was considered state-of-the-art. Cray supercomputers in the late 1980s had internal "clock" speeds of 4 nanoseconds (250 MHz) and were once restricted for sale to foreign governments. Today, similar processors cost under $5,000 and can be bought at many stores in the US. Even access to sensitive GPS satellite information, once reserved for the military during threat conditions, will no longer be "scrambled," and will instead be opened for use by the public—and to foreign governments, as well.

The upshot is that other nations need no longer make large investments in R&D if they have access to products generated from the U.S. R&D infrastructure. In order to produce new weapons based on advances in science and technology, other nations simply have to wait for the United States to provide them access to the new technology.

Our nation's response to date to this threat is to ensure that it doesn't stop making advances. This doesn't mean that the United States should be performing military-related R&D for the rest of the world while they ride our coattails in technological breakthroughs. Rather, the United States is making the R&D advances to support the third of the three pillars of our national power—the strength of our economy—and marketing the advances commercially. As a result, the technological edge that the United States once commanded—the argument of quality over quantity—can only be maintained if a *continuing investment* is made to ensure our technological superiority.

A drawback to this strategy is in the way that other countries view technology transfer. For example, it has been said that the United States treats technology as a *commodity*—we will sell technology, and in effect the fruits of our R&D, to make a profit. On the other hand, countries such as Japan view technology as a *national asset*—it keeps its technology and instead sells products based on that technology. Which strategy is best? One can look at the growing strength of Japan's economy and the subsequent trade imbalance with the United States to draw one's own conclusion.[31]

**Provide technical solutions to achieve future joint warfighting capabilities.** Admiral William A. Owens, Vice Chairman of the Joint Chiefs of Staff, said, "Technology will never be a substitute for courage or human touchness in conflict, but it will increase the likelihood that the tough and courageous will be successful."[32] To do this, *it is imperative that the technology created in the laboratory be efficiently transitioned to the warfighter.* The best technological advance in history is of no use to the person fighting America's war if that discovery is not made available in a useful way. For example, Ernico Fermi's ground-breaking validation of the nuclear chain reaction process was of no value to the soldiers in World War II while Fermi's laboratory experiment remained under a football field at the University of Chicago. It was only through the advanced development program conducted by the Manhattan Project at Los Alamos—the atomic bomb project—that gave the United States its first nuclear weapons.

- *Efficiently Transferring Technology.* Future joint warfighting capabilities are defined as a requirement by the warfighter, the operational arm that is on the "front line" of our country's battles. The concepts that evolve into the specific programs that satisfy these capabilities are supported by science and technology programs. Basic research (6.1) provides the foundation; exploratory (6.2) and advanced development (6.3A) builds on this foundation.

  The S&T accomplished for the warfighter feeds into joint technology programs called technology demonstrations. These demonstrations are designed to demonstrate that the concept developed during the S&T phase satisfies a militarily significant goal. By increasing the level and the scope of the technology demonstration, some of the practical problems that always arise during operational use are ironed out.

  In the next section a specific program (Advanced Concept Technology Demonstrations) is discussed that accomplishes part of this goal. Irrespective of the specific program, the people participating in defense-related science and technology must work with the warfighters at every opportunity throughout the R&D phase. *They must understand the warfighter's needs, and they must move their most promising concepts rapidly to satisfy the warfighter's requirements.* This is accomplished only by keeping in close, continuous contact with the warfighter and keeping in focus the ultimate purpose for conducting defense-related S&T: to win wars.

Note that this ultimate purpose supports the first of the three "pillars" (capability) that serve as the foundation for our national power: ensuring the readiness and capabilities of our military forces, engaging with other nations to prevent conflict from occurring, and strengthening our economy. It may be argued that these three pillars are merely stopping points along the path of winning a war, or even that the *real* purpose of defense-related S&T is to "develop options for decisive military capabilities."[33] However, it must be remembered that just as the commercial marketplace has as its bottom line the profit of its endeavor, in the same respect—but more powerfully, for the survival of our nation may depend on its degree—the bottom line, the ultimate purpose of defense-related S&T, is winning that war. No matter what policy makers may say, if they subscribe to any other result, than it is *not* the warfighter they are supporting, but rather a beguiling veil of hidden political agendas.

- *Avoiding Technology Surprise.* Finally, the necessity of providing technical solutions to achieve the future joint warfighting capabilities serves to avoid technological surprise. Scientists and engineers must be fully aware of the scientific advances made throughout the world and keep abreast of new technologies. The advances may not only help to solve a localized problem, but it may also shed light on what new military application is being considered for use against our forces.

**Balance basic research and applied technology in pursuing technological advances.** The National Science and Technology

Strategy commits the United States to expanding "fundamental scientific knowledge that may lead to future warfighting capabilities."[34] Furthermore, DOD acknowledges that the technical superiority that leads to these future warfighting capabilities is grounded in scientific advancements and thus broadly invests in "defense relevant scientific fields." As such, the objective of DOD investments in basic research is first to discover new knowledge and second to invest in and support a community of world-class scientists as they research the latest and technically superior weapons of war.[35]

By its very definition, basic research provides the foundation for both military and non-military applied research. Because basic research is not directed toward specific goals, the DOD basic research program also supports the strength of our economy. This then strengthens our economic security as well as our national security.

DOD has four investment strategies in maintaining the quality and excellence of its science and technology programs:

- *Support Quality Basic Research.* Only the very best scientists are able to conduct the critical, cutting edge research that advances the state-of-the-art in science and technology. These scientists must be able to recognize potential applications and fully understand the ramifications of far-reaching, and often esoteric discoveries. For example, an arcane advance in theoretical electromagnetics made in the former Soviet Union was recognized by a mathematician working for a defense contractor as being applicable to negating the return of radar signals from a specially shaped body. Out of this discovery grew the stealth concept, and within *months* a hastily cobbled together experiment (the Air Force's *Have Blue* program) proved the concept and launched one of the most revolutionary technological advances seen in this decade, the F-117A fighter.[36]

DOD's goal is to support first-rate scientists in universities (where traditionally over 50 percent of the basic research is performed), industry, defense laboratories, and other governmental laboratories such as those managed by the Department of Energy, Department of Commerce, and the national security agencies. Through its program of providing "6.1" funding as competitive grants, DOD is committed to supporting the best researchers based on merit, regardless of organization or location. [37]

- *Sustain Stable Research Funding.* Basic and applied research is not faucet to be turned on and off. The very best research is performed only when given the ability to come to fruition after many years of work. It is easy to make mistakes when new ground is pioneered, and in order to follow the scientific method of obtaining reproducible results, much care must be given to sometimes subtle changes in data. Careful consideration of data takes a long time, and funding must be given to ensure the correct results are obtained.

Modern research is conducted in teams. A perusal of an esteemed publication such as *Physical Review Letters*, arguably the top peer-reviewed journal in physics, shows that research teams consist from three to over a hundred people. The larger teams are consolidated around "big science" projects, such as Tokomaks for magnetic fusion, or large particle accelerators for work in elementary particle physics. The point is that all these teams revolve around a few key individuals, who cannot be easily replaced. If team leaders and a significant number of members leave the teams because of layoffs, or even budget uncertainties, then the project is not just put on hold until another team can be assembled but usually scrapped, sometimes for many years. These highly trained individuals are true national assets, not interchangeable parts.

The reason for this situation is grounded in economics. There is an extremely small supply of top scientific talent (Noble laureate class), and if the demand for that talent is large, then the talent will pursue other activities. Finding key individuals to direct world-class basic and applied research is *not* the same as massively hiring engineers. The number of talented individuals on a scientific team does not decrease linearly as funding decreases; rather, there is a sudden drop to zero if a threshold is reached and the funding drops below a certain level.[38] Therefore, the ability to sustain research funding at a stable level is critical for keeping key scientists, ensuring continuity, and ultimately ensuring the research remains world class.

- *Educate future scientists and engineers.* The United States made a long-term commitment in the 1800s to support agriculture and provide economic incentives for continued agricultural growth. As a result, the United States saw unparalleled growth in farming, and is now the world leader in producing food supplies and sustaining agricultural growth. In much the same manner, if the United States is serious about sustaining our technological lead in forging the foundation for military strength, it must make a long-term commitment to motivate and stimulate the next generation of scientists and engineers.

Today the United States has the world's best universities for producing future scientists and engineers. A recent study conducted by the National Academy of Engineering concluded, "The superiority of advanced technical education in America is recognized throughout the world, as attested by persistently large enrollments of foreign students in doctoral level engineering and science programs in American universities."[39] As a result, American universities command a large percentage of the nation's total R&D budget and a majority of the foreign students educated in America remain due to the enhanced economic conditions as compared to their homes of birth. However, because of the poor state of America's primary and secondary education system, the number of U.S. citizens graduating with advanced technical degrees has been *decreasing* throughout the years.[40] Therefore, although American universities are *the* best in the world, we are also decreasing the number of our own science and engineering graduates. This is contrary to what existed in the 1960s when our schools were not only the best, but we also had a pool of competent secondary education graduates which went on in record numbers to become technically trained individuals.

One reason the aerospace industry underwent explosive growth in the 1960s was because of the U.S. response to Sputnik and the spinoff opportunities presented by the rise in new jet airplanes.[41] Universities pumped out engineering graduates to meet the demand for ever increasing numbers of engineers, and times were heady for new scientific talent.[42] During the same period, the defense industry also attracted young scientists and engineers, as new weapons of war demanded ever-increasing scientific acumen. However, the huge number of layoffs during the economic downturn of the early 1970s of science and engineering workers showed people that the science and engineering field was not an infinite source of jobs.

Another example of drought where scientific talent once reigned is in the rocket engine and nuclear weapons industry. In the 1960s American rocket engineers built the largest and most powerful rocket known to man—the F-1 liquid oxygen kerosene-fueled engine that powered the giant Saturn V launch vehicle.[43] The F-1 engine produced over 1.5 million pounds of thrust and provided the massive first-stage lift capability to launch a moon vehicle and return capsule to orbit. However, because the Apollo program was scuttled and the F-1 manufacturing line was stopped, the infrastructure to build the F-1—the workers, machines and even the blueprints—are no longer around. It is estimated that it would take 10 years and over a billion dollars just to rebuild to the same level achieved by our engineers 30 years ago. If cheap access to space is ever realized, perhaps through a true single-stage-to-orbit vehicle, then this era may have just served as a stepping stone; but if heavy lift rockets will again be built, then we will have to relearn this critical technology. Therefore, a key question is whether or not this expertise is even needed.

The nuclear weapons industry is quickly following suit. Established as the Manhattan Project and having produced an unparalleled technical achievement that won World War II, the infrastructure that maintained the atomic weapons industry has ground to a halt. With the threat of a massive nuclear attack from nuclear weapons abating worldwide, the United States is moving out of the nuclear weapons business. As a result, the number of experienced scientific talent that designed, validated and nurtured the weapons has been decreasing throughout the years. The average age of the nuclear researcher has increased, and little new talent has been brought in to act as scientific guardian of the nuclear genie.[44] As a result, the dominance of nuclear weapon S&T once held by the United States is in severe danger of disappearing. And again, the key question is do we really need to maintain this dwindling expertise, or will new technologies demand us to train scientists in other areas of expertise?

With these analogies directly applicable to our present situation, if we are going to continue our dominance in defense-related S&T, we *must* invest in our future by educating the highest quality people to become scientists and engineers. Moreover, it is not enough merely to provide high-quality education al opportunities for these people; they must also be *motivated*, through the use of various incentives, to follow a career path in defense-related research.

The John and Fanny Hertz Foundation had a vision for providing such motivation for high-quality undergraduates. The foundation established a set of highly competitive scholarships purposely designed to bring the best and the brightest undergraduate science and engineering students into the competitive arena. The prize—a full scholarship with stipend to some of the best universities in the nation, with the motivation to study in the applied sciences. The scholarship recipients are funded to work all the way to their Ph.D. and are fully cognizant of the fact that they may be exposed to national security-related S&T. The Hertz Scholarship is a shining example of motivating our future scientists by exposing them to a favorable view of defense-related S&T. Graduates of this program include Dr. Peter Hagelstein, inventor of the X-ray laser, among several other luminaries.[45]

Another vision for producing scientists and engineers motivated to work on defense-related research is through the use of the scientific equivalent of "land grant" colleges. Established in the 1890s, the agricultural (A&M) state universities were responsible for an explosion of agricultural advances. In the same manner, nationally supported schools focusing on the applied sciences and engineering, might someday produce graduates that have been exposed to unique defense-related requirements. And similarly, as in the A&M schools where the graduates are not forced to go into agricultural-related jobs, there should be no mandatory requirement for these graduates to go into defense jobs. The point is to produce self-motivated individuals. Just as it would be foolish to try and establish an engineering school to compete with the MITs of the world, scholarships and other directed programs might prove to be the acceptable path rather than building new buildings and campuses.

- *Promote teamwork and partnerships.* The era of defense-related laboratories maintaining sprawling research facilities to cover every aspect of research are over. The Department of Energy laboratories did this very well: they performed the basic and applied research of nuclear weapon development (such as atomic physics), and the system took this all the way to producing a working, operational warhead. However, the nation was mobilized during the Cold War to perform this activity, and cost was a secondary (if even that high) factor in maintaining the infrastructure needed to undertake this enormously complex endeavor. And it was accomplished because the cost was worth it compared to the threat. Today, according to national policy makers, that is no longer true.

Therefore, to ensure that the highest quality of scientific work is performed, the DOD research program envisions partnerships of defense laboratory researchers with their colleagues at Universities, industry, and other national laboratories. The teams would be drawn together to use common facilities, and would exploit the strengths of various team members—a national laboratory may be particularly strong in large scale supercomputer simulations, while several university teams might be able to provide the analytic support, and a DOD laboratory may well have the proper equipment to perform a crucial experiment to validate the theoretical results.

There is a parallel to this in American industry. A shift occurred in the commercial market place when corporations downsized during the 1980s and 1990s. The corporations concentrated their energy in their core competencies—the things they did best. What this meant was that instead of being able to be fully independent, and existing in isolation from other companies by possessing everything internal to their organization—artwork, film processing, technical editors, messengers, machine shop, etc.—the companies cut to the core of what they did best, their "core competencies," and contracted out everything else.[46]

In the same manner, defense laboratories must focus on their core competencies and do what they do best. They must exploit the expertise of others in getting their product—the technology for new weapons of war—out the door and to the warfighters. The downside is that those laboratories who have chosen the wrong core competency—either through politics or through internal agendas—will quickly discover that they cannot compete in this new "marketplace" of defense-related S&T.

As an example, consider a laboratory that excels in scientific field "A." If, because of politics, that laboratory declares it is now competent in field "B" and drops "A," then no matter how good it thinks it can perform "B," it is still second rate. It has not correctly chosen its core competency, and when cuts are made, this lab will be ripe for cutting.

**Incorporating affordability as a design parameter.** The Federal Government maintains over 700 federal laboratories with an annual budget of over $20 billion, nearly one third of the $70 billion spent yearly on S&T in the U.S.[47] Of this amount, the 1996 fiscal year (FY96) defense science and technology budget totals $7.8 billion, broken down as shown in the table below.

The remainder of the national security science and technology budget is spent by Department of Energy (DOE) and other national security laboratories. In the 1980s, the "big three" DOE weapons laboratories—Los Alamos, Lawrence Livermore, and Sandia National Laboratories—had budgets in excess of $1 billion a year and employed on the order of 7,500 full-time equivalents (FTEs) apiece, not including contractor personnel that substantially added to the manpower. In FY 1995, the DOE Defense Programs involved

59,794 FTEs, including 7,321 at Lawrence Livermore, 6,960 at Los Alamos, and 8,495 at Sandia National Laboratory, excluding contractors.[48] Note that the substantially greater number of FTEs at Sandia includes both Sandia's Livermore (mostly energy-related) labs as well as the bigger Albuquerque labs. And although all three of the DOE weapons laboratories conduct national security science and technology as well as research and development, Los Alamos is generally considered more of a "basic research" lab, Lawrence Livermore an "applied research" lab, and Sandia an "engineering" lab.

FY1996 Defense Science and Technology Budget

| | |
|---|---|
| Air Force | $1.4 B |
| Army | $1.1 B |
| Navy | $1.4 B |
| Advanced Research Projects Agency | $2.6 B |
| Washington Headquarters Services | $0.7 B |
| Defense Nuclear Agency | $0.2 B |
| Other Defense Agencies | $0.4 B |
| Total | $7.8 B |

Source: Director, Defense Research and Engineering.

The Department of Defense procurement budget has dropped by two thirds from 1985 to 1996, with the science and technology budget remaining relatively constant over the same period.[49] Although this does not take into account the one-third drop in S&T budget from the mid-1960s, the fact is, money is scarce.

As a result, DOD has established that is critical for S&T programs to adopt cost-cutting measures in current and future weapon and support systems and that this should be a primary objective in for all acquisition activities. As nearly 80 percent of the life-cycle cost of a weapon system is determined during the concept and preliminary design phase (corresponding to approximately the first five percent of the life cycle cost), it is even more critical than before that cost-cutting measures be adopted as a design objective.[50] The DDR&E has initiated several cost-cutting measures and is actively championing the services to institute them.

Because the present funding situation drives every S&T decision, it is useful to explore how this affects DOD S&T strategy. Below are six major cost-cutting thrusts deemed critical by the Clinton administration.

- *Use the best commercial products, practices and capabilities.* This involves purposely moving away from defense-unique "mil-spec" standards and embracing national and international commercial market-driven standards and practices. DOD has formally embraced this policy and has made the use of mil-spec standards the *exception* rather than the rule. Flexibility and exploitation of OTS (Off-the-Shelf) technologies will force DOD away from developing unique capabilities that has the inherent danger of driving cost beyond reason.

  Horror stories abound about the inordinate cost of "mil spec" tissues: the length of time needed to get it in stock, and the harsh texture, in comparison to the availability of commercially available tissues. In his effort to Reinvent Government, Vice President Gore coined the phrase, "If it makes sense, and it's not illegal, immoral or unethical, and the cost benefit is there, then just do it." Procuring commercial tissues is the quintessential example for getting away from mil-spec and doing what "makes sense." This is but one example of the savings realized by adopting commercial standards and practices, and S&T research must strive to make similar savings.

  However, DOD does not procure only tissues. On weapons systems there exist the possibility that the military requirement for a critical capability might be degraded if "exceptions" are not made. For example, if a logic chip is needed for a satellite that has a mission to survive a nuclear EMP blast, then it does not make sense to use a commercially available chip that has not been tested in that environment. This then requires that the concept be thoroughly "shaken out" at the requirements stage to ensure that the requirement is really needed, and if so, *then* the exception can be made.

One way to look at this is by using the concept of marginal returns. For example, is a 10 percent improvement in capability worth the investment? And if not, are the requirements still satisfied? It is necessary to understand what capabilities drive the cost of the system—and if those capabilities are really needed.

- *Simulation and modeling.* Advances in computers, software, operating systems, algorithms, storage media, and associated multimedia devices have elevated the field of simulation and modeling to heretofore unachievable heights. Modeling is generally considered a subset of simulation—a *model* may involve a single process which may or may not be time dependent, such as a model of a chemical reaction; a *simulation* consists of a time-evolving system of several models, usually all interacting and interconnected.

Not only is it possible to simulate how new technologies may be used to help the warfighter, but it is also possible to evaluate a wide spectrum of consequences brought about by the introduction of that technology to the battlefield. All this and more can be accomplished in the early phase of science and technology development, before costly decisions are made in the acquisition process. Simulation can feed into producing better requirements, improved cost-benefit analysis, comprehensive tradeoff analysis, optimized designs, and improved test and evaluations.

Simulations can be used in all phases of the acquisition process, with particularly strong savings at the beginning of the

R&D cycle. Simulations provide the means to conduct an extraordinary amount of "numerical experiments" to hone down the parameter space of possible paths. For example, at the basic research level, *ab initio* simulations of the quantum interaction of light with matter yields insight into nonlinear phenomenon, resulting in new theoretical models at a macroscopic level; at the exploratory development stage, simulations of radiation-matter coupling validates the macroscopic theories and may point to a application; parameter studies may then be conducted to optimize initial engineering designs at the advanced development level.

- *Improve Manufacturing Processes.* Manufacturing in the U.S. accounts for over 21 million jobs and nearly one-fifth of the gross domestic product, totaling over $1 trillion dollars a year.[51] In addition, the manufacturing sector employees over 75 percent of the U.S. scientists and engineers and conducts nearly 90 percent of the nation's non-defense private R&D.[52] It is in the government's best interest not to hinder this immensely successful endeavor.

Manufacturing in the United States has undergone a major change in the last two decades. Rising to the threat of international competition and, a decaying infrastructure, and fueled by investments made by DOD in the 1980s, manufacturing is rapidly instituting changes to reduce cost, enhance quality, and add new capability to make it more agile and flexible.

In the past, economies of scale enabled costs in manufacturing to be held low. With today's shift away from keeping large inventories and moving toward a "just in time" delivery system, manufacturing runs are becoming smaller and more unique. As a result, an effort is being made to institute new technology so that it is easy to reconfigure manufacturing lines. This will allow smaller, more economical, and variable-volume manufacturing runs.[53] Some example of these new technologies include:[54]

-*Smart materials*—Materials that incorporate sensors and actuators that may dynamically respond to stresses, temperatures, electromagnetic radiation and the like.

-*Aggregation processes*—Includes the ability to produce items close to the final production form. For example, in stereolithography, prototype parts are made "directly" from computer files using droplet jet-spraying, chemical vapor deposition, sputtering and other processes.

-*Microelectromechanical systems and molecular manufacturing*—Microelectronic devices (at the micron scale and below) are being invented using microfabrication and micromachining processes. A recent study by the Office of Technology Assessment, *Miniaturization Technologies*, forecasts the first version of molecular-scale devices within the next 10 years.[55]

Other more conventional advances such as numerical manufacturing allow metrics to be established. This enables low production runs, reduced test and evaluation phases, and earlier detection and correction of manufacturing errors.

- *Consider environmental factors*. In the past little concern was given to the environment and as a result, the United States was faced with unprecedented number of environmental disasters, such as DOE's Rocky Arsenal debacle. DOD has had its share of environmental problems, including the presence of PCBs and jet fuel either accidentally spilled or purposely dumped onto the ground on military bases. Like DOE, DOD has been forced to clean deal with these issues and clean up the pollution they might have inflicted on the environment. For example, in 1992, for the first time in DOE the environmental cleanup budget actually exceeded the nuclear weapons research budget. Because of these huge increases in cost and subsequent possibilities for inflicting life threatening damage, life cycle costs must reflect a true estimate of environmental costs, including pollution prevention and cleanup. Advances in science and technology may help bring down these costs, and if the investment is made up front, then these issues can be addressed before the costs spiral out of hand.

- *Establish service affordability programs.* Each of the four services—Air Force, Army, Marines, and Navy—has instituted S&T programs that address affordability issues. Although no details are given for specific programs, it is the administration's policy that the services will "integrate and strengthened these into a coherent program that emphasizes Service-unique needs and addresses the broadest affordability challenges."[56]
- *Reduce the Cost of Ownership.* DOD is charged by administration policy to vigorously investigate new applications of science and technology that can reduce the cost of operating, maintaining, and upgrading systems.[57] Some examples of new technologies include developing "smart skins" that can use self-diagnostic techniques for determining damage, corrosion and fracture sensors, and various non-destructive testing techniques, such as neutron bombardment. Included in this task of reducing the cost of ownership is the implementation of "smart" logistical practices. This includes focusing on rapid turn around and providing a balance between invoking "just in time" delivery systems and maintaining vast warehouses of inventoried material. The details of this argument is beyond the scope of this book, but it brings to light one of the major debates in providing "smart" logistical practices.

## DEFENSE STRATEGY IN ACTION

Think of it as evolution in action.

Niven and Pournelle, *Oath of Fealty*

### PARTNERING AND EXPLOITING EXPERTISE

The Department of Defense no longer drives R&D in the United States. Industrial investment in U.S. military R&D has decreased from a high in 1985, with the trend in all military relevant R&D (including non-military S&T, industry S&T, IR&D, and military S&T) decreasing from $18 billion in 1986 to just under $12 billion in 1996, a drop of one-third.[58] With this downward trend, it is incumbent on the U.S. Government to "smartly use" its remaining R&D resources.

The President's charge to "Develop and transition superior technology to enable affordable, decisive military capability and to enhance economic security," clearly has as its bottom line to improve productivity and reduce costs by working with commercial industry.[59] The present administration has chosen to do this in three ways: 1) leveraging funds; 2) increasing the level of knowledge in the industrial sector; and 3) diffusing new technology.[60]

"Leveraging funds" means to exploit areas of excellence of R&D either in the public or private sector; it also means reducing any redundancies that may exist, especially in the public arena. For example, it is widely believed that DOD should not be performing the same research as the private sector, such as in computer chips. *Rather it is up to DOD to focus its research on those areas that have no commercial market, yet underlies defense S&T*. In the past, up to 40 percent of military R&D was conducted by defense or DOE laboratories; this ratio is expected to decrease as the commercial marketplace is "leveraged."[61]

Specific vehicles such as CRDAs (Cooperative Research and Development Agreements) with industry is one way to "leverage" the commercial marketplace by transferring technology from government labs to the private sector. Another vehicle, the Technology Reinvestment Program (TRP), was established to integrate defense and commercial production facilities, deploy technology to and from commercial industries, develop dual-use technologies, and diversify from defense to commercial products. In FY94 the TRP received $404 million out of a total $1.7 billion total targeted for defense conversion and dual-use technology.[62] However, the TRP concentrated on developing technology and *not* on transitioning basic research to a higher development stage. In addition, many questions remain as to how the technology areas are selected. The next chapter will return to this question and illuminate other concerns.

"Increasing the level of knowledge in the industrial sector" is the second key to reducing costs and increasing productivity. This will be accomplished when DOD and industry understand defense technology needs. Much work has been done in this area, and the *DOD Key Technologies Plan*, released in 1992 and tied to the defense S&T Strategy, marked the beginning of defining DOD technology needs. The present needs are grounded in national S&T security strategy and flow from the JROC future warfighting capabilities, as shown at the beginning of this chapter.

"Diffusing new technology" is the final key for reducing costs and increasing productivity. Again, much work has been accomplished in this area, and programs such as CRDAs are but one means for diffusing technology. Government-commercial consortia such as the Great Lakes Composites Consortium or the Army's National Automotive Center address sectorwide capabilities.[63] Broad area concerns are addressed that affect both defense and commercial needs. Program-specific actions such as the DOD Manufacturing Technology program (MANTECH) or the Manufacturing Operations Development and Integration

Laboratories (MODILs) promote process integration at the industrial-sector, firm and facility levels.[64]

In general, a wide-area, broad-based strategy would best enhance both defense and commercial industry. A DOD advisory group on materials processing concluded in 1993 that "the most important government roles in advancing the technologies of manufacturing systems are to provide seed money for promising technical opportunities that would not otherwise be pursued and to bring individual companies together for mutual leveraging in areas of common need."[65] The group found that programs aimed at enhancing a particular weapon system or company would not permeate throughout industry—the risks are too high, the development costs are to costly, and the payback time is too long to take a chance on any one system.

## TRANSITIONING FROM BASIC RESEARCH: ADVANCED CONCEPT TECHNOLOGY DEMONSTRATIONS (ACDTS)

There is an absolute necessity for the scientist and engineer to be in constant communication with the warfighter throughout the R&D phase. The warfighter can provide the requirements for the eventual capabilities that the weapon may have, and therefore indirectly influence the direction of technological advance. Close communication between the scientist and warfighter will thus ensure that *technological opportunity* will be matched with the warfighter's required capability.

A program has been developed in DOD to exploit such an exchange of information between the warfighter and the technologist. Called the ACTD (Advanced Concept Technology Demonstration), this program allows the warfighter to "experiment in the field with new technology in order to evaluate potential changes to doctrine, operational concepts, tactics, modernization plans, and training."[66]

The administration fully endorses the DOD ACTD program for "capturing and harnessing innovation for military use rapidly and at a reduced cost."[67] ACTDs are focused on four principal objectives:

- Gaining an understanding of and evaluating the military utility of new technological applications before committing to an acquisition;
- Developing Concepts of Operations and new doctrine that best makes use of the new capability;
- Providing residual operational capability to the forces;
- Improving the acquisition decision process.[68]

The intent of the ACTD program is to provide the user with detailed interactions with the proposed weapon system very early in the development cycle. Typical ACDTs last from 2 to 4 years, thus providing a means to rapidly and cost-effectively introduce new capabilities to the warfighter. If successful, ACTDs are then transferred to one of the military services or a defense agency for acquisition.

## SUMMARY

Defense science and technology strategy is ultimately grounded in the national security goals of the United States. The Joint Requirements Oversight Council (JROC) has determined five future warfighting capabilities that will satisfy the national security goals. From these projected capabilities spring four requirements set by DDR&E, the Director of Defense Research and Engineering:

- Maintain technological superiority in warfighting equipment
- Provide technical solutions to achieve future joint warfighting capabilities
- Balance basic research and applied technology in pursuing technological advances

- Incorporate affordability as a design parameter.

The defense science and technology strategy "clothes" these requirements to ensure that the mainstay of the Administration's policy—improving productivity and reducing costs by working with commercial industry—is first and foremost on the minds of DOD. This then concludes the discussion of the present defense research strategy. The next chapter highlights specific concerns not addressed by the strategy and points out some flaws in its logic.

## NOTES

1. Director of Defense Research and Engineering, *Defense Science and Technology Strategy* (Washington, DC: USGPO, September, 1994), 1.

2. Ibid., 3.

3. Ibid., 4.

4. Ibid., 6.

5. Ibid., 5.

6. Federal Express has optimized using operational analysis techniques for quickly and efficiently calculating the best routes; these routes are changed dynamically to ensure market competitiveness.

7. Director of Defense Research and Engineering, *Defense Science*, 4.

8. Ibid., 5.

9. Ibid., 5-6

10. Ibid., 6.

11. The Los Alamos National Laboratory has proposed using ozone to mitigate the threat of biological weapons; similarly, the Air Force's Phillips Laboratory has proposed using high energy X-rays.

12. Aerospace Industries Association of America, Inc., *Aerospace Facts & Figures: 1995-1996* (Washington, DC: Aerospace Industries Association, 1995), 63.

13. Ibid.

14. Orbital Sciences Corporation has successfully launched payloads to low earth orbit with its *Pegasus* booster, a rocket attached underwing to a DC-10. In the summer of 1995 a cadet from the USAF Academy majoring in astronautical engineering spent his summer performing research for a Virginia-based company building a 10-pound satellite under Air Force Space Command sponsorship.

15. Air Force Science Advisory Board, *New World Vistas: Air and Space Power for the 21st Century* (Washington, DC: HQ USAF/SAB, 1995), 42-46.

16. Jacques S. Gansler, *Defense Conversion: Transforming the Arsenal of Democracy* (Cambridge, MA: MIT Press, 1995), 2.

17. With the recent concern of nuclear weapon proliferation, one can open almost any science text and find an elementary explanation of how to build an atomic bomb. (I still have my own notes from a 7th grade physics class.) Much of the theory behind these weapons is known in the open literature. For example, Grolier's *New Book of Knowledge* 13 N, 363, gives an explanation of how to make an atomic bomb—but yet my 2nd grade daughter routinely uses Grolier, and has no desire to proliferate nuclear weapons.

18. See, for example, Kevin J. Anderson and Doug Beason, *The Trinity Paradox* (New York: Bantam, 1992), which was a finalist for the American Physical Society's Forum Award for explaining science's role to the general public.

19. As mentioned in an above footnote, the Los Alamos National Laboratory had an advanced technology proposal to investigate the use of concentrated ozone for mitigating biological and chemical weapons. Studies shown that the ozone may "quench" the weapons, as opposed to destroying them. The Air Force's Phillips Laboratory has proposed using bursts of high energy X-rays to actually destroy biological weapons in flight, while they are still in the warhead; however, extremely high dose rates are needed for this concept to work, and problems remain such as material "shadowing," or blocking the X-ray effects.

20. Executive Office of the President, Office of Science and Technology Policy, *National Security Science and Technology Strategy* (Washington, DC: U.S. Government Printing Office, 1995), ii.

21 Committee for National Security, National Science and Technology Council, The Office of Science and Technology Policy, The White House, *National Security Science and Technology*, cover letter by President Bill Clinton (Washington, DC: USGPO, 1995).

22 Gregory H. Canavan and John F. Lilley, "Managing Public Sector Research and Development: Innovation Versus Responsiveness," in *Defense Technology*, A. Clark IV and J. Lilley, eds. (New York: Praeger, 1989), 244.

23. Gansler, 10, 169-170, 199-218.

24. Canavan and Lilley, 244.

25. Gansler, 10.

26. Director of Defense Research and Engineering, *Defense Science*, 1.

27. There may some unusual circumstances where a system may be developed and skip the production phase. This may occur when only one or two

systems are fielded, or if the system is needed in an urgent manner. For example, the "buster bunker" kinetic energy weapon was rushed from the laboratory to the field during Operation Desert Storm, bypassing the normal development phase. Another example is the identification system for ground vehicles (determining if they were friend or foe) that was developed by the Defense Advanced Research and Development Agency (DARPA)in 16 days during *Desert Storm*, only possible because of DARPA bypassing all the standard DOD reporting procedures.

28. Director of Defense Research and Engineering, *Defense Science*, 2.

29. Deputy Director of Defense Research and Engineering, *Defense Science and Technology (S&T) Strategy*, unpublished briefing given to Congressman Dicks and his personal staff, December, 1995.

30. Donald L. Losman and Shu-Jan Liang, *The Promise of American Industry* (New York: Quorum Books, 1990), 92.

31. During ICAF's Space Industry Study trip to Japan in April 1996, I spoke with a State department employee about this issue of Japan treating technology as an asset. There is much concern throughout the United Stats that it is giving away the fruits of its R&D, while nations such as Japan treat technology more as the foundation for business, and not as a mere commodity.

32. Stuart E. Johnson and Martin C. Libicki, eds.,, *Dominant Battlespace Knowledge: The Winning Edge* (Washington, D.C.: National Defense University press, 1995), 14.

33. Director of Defense Research and Engineering, *Defense Science*, 1.

34. Ibid., 33.

35. Ibid., 33.

36. Ben Rich and Leo Janos, *Skunk Works: A Personal Memoir of My Years at Lockheed* (New York: Little, Brown and Company, 1994).

37. Director of Defense Research and Engineering, *Defense Science*, 33.

38. In Physics this is called a "Fermi function."

39. National Academy of Engineering, *Prospering in a Global Economy: National Interests in an Age of Global Technology*, Thomas H. Lee and Proctor P. Reid, eds. (Washington, DC: National Academy Press, 1991), 57.

40. Ibid., 62.

41. The unprecedented growth of the airline industry in the early 1990s, where total aerospace industry sales peaked in 1991 at $139 billion—two and a half *times* greater than 1980—was primarily due to deregulation, which intensely increased competition. This is different from the growth experienced in the 1960s, which has its roots in the overall growth of science and technology, mostly attributed to the Sputnik "threat."

42. The last comparable period of spiraling scientific and technological growth was during the days of the Strategic Defense Initiative, when the nation's S&T community was aligned either for, or against this ambitious program.

43. Astronaut General (USAF, ret.) Thomas P. Stafford, The *Synthesis Group* report to the Vice President: *America at the Threshold: America's Space Exploration Initiative* (Washington D.C.: USGPO, 1991), 65.

44. From 1990 to 1995 I had several discussions with key nuclear weapon designers both at Los Alamos (X-Division) and Lawrence Livermore National Laboratories (A and B Divisions) that telegraphed their worry that "new talent" was not being introduced to the nuclear weapon business. It is getting hard to attract, and to keep, top scientific talent to perform vital research in what is perceived as a dying field. In 1996 the Department of Energy, in concert with the Department of Defense, established a nuclear "dual revalidation" program to revalidate the weapons in the nuclear stockpile. This process may serve to attract some new scientific talent, but it remains to be seen what the long-term health of this arcane interdisciplinary field will be.

45. William Broad, *Star Warriors: A Penetrating Look Into the Lifes of the Young Scientists Behind our Space Age Weaponry* (New York: Simon and Schuster, 1985), 44.

46. Dr. John White, Deputy Secretary of Defense, at a lecture to the Industrial College of the Armed Forces, National Defense University, on February 8, 1996.

47. Gansler, 112.

48. These data were obtained from the U.S. Department of Energy personnel office, 1995.

49. Director of Defense Research and Engineering, *Defense Science*, 25.

50. Ibid., 25.

51. *Science and Technology*, A Report of the President transmitted to the Congress 1993, White House Office of Science and Technology Policy (Washington DC: USGPO, 1993), 161.

52. Ibid., 161.

53. Director of Defense Research and Engineering, *Defense Science*, 26.

54. *Science and Technology*, A Report of the President transmitted to the Congress 1993, 161-181.

55. Ibid., 170.

56. Director of Defense Research and Engineering, *Defense Science*, 26.

57. Ibid., 27.

58. Deputy Director of Defense Research and Engineering, *Defense Science*.

59. Director of Defense Research and Engineering, *Defense Science*, 2.

60. U.S. Congress, Office of Technology Assessment, *Assessing the Potential for Civil-Military Integration: Technologies, Processes, and Practices.* OTA-ISS-611 (Washington, DC: U.S. Government Printing Office, September, 1994), 116.

61. Ibid., 116.

62. Ibid., 117.

63. Ibid., 121.

64. Ibid., 121.

65. Ibid., 122.

66. Director of Defense Research and Engineering, *Defense Science,* 24.

67. Executive Office of the President, Office of Science and Technology Policy, *National Security Science and Technology Strategy* (Washington,DC: U.S. Government Printing Office, 1995), 20.

68. Ibid., 21.

# 5. KINKS IN THE STRATEGY

## THE PROMISE OF THE COMMERCIAL MARKETPLACE

U.S. national security is dependent not only on the strength of our military forces, but also of our economy.[1] As such, it can be argued that our economic growth and strength have been somewhat restrained because of the past expenditures on military systems, diverting resources from commercial industry to the military. For example, in 1988 over 65 percent of federal R&D outlays were allocated to defense—capital, personnel, materials—compared to less than 0.2 percent for industrial development.[2]

However, it can *also* be argued that through spinoffs, R&D investments, re-investment, and infrastructure building, the effects of diverting resources to military needs have not been as bad as some critics claim.[3] Whatever forces shaped our country's economic growth in the past, however, we must live with less defense spending than in the past and draw on the commercial marketplace to generate key technologies.[4]

It is the belief of some that the majority of R&D should be conducted in the private sector rather than in government laboratories. They argue that this will exploit the efficiencies brought about by the competitive nature of the marketplace. In addition, a close coupling exists between commercial R&D and the marketplace, which further drives these efficiencies.[5]

The technological superiority of high performance military systems must be obtained at a reasonable cost. It is the administration's policy to "use the same technology and the same industrial base, where appropriate, to build military and commercial products."[6] Chapter 3 detailed several strategies to strengthen the commercial-military industrial base. These strategies included developing dual use technologies and processes, sustaining investment in high priority technologies, exploiting commercial technologies, strengthening technology transfer, and fielding Advanced Concept Technology Demonstrations. It is through these strategies that the U.S. government intends to support the foundation for future weapon systems. Perhaps the quintessential example of such a strategy is that of what was done with military specifications.

In 1994 when the Department of Defense reversed it's long-standing policy of using only military specifications ("milspecs") for weapon systems, over 31,000 detailed specification standards were changed overnight.[7] The military services were directed to use commercial (or performance based) specifications in lieu of milspecs "unless no practical alternative exists." This change has dramatically changed the way defense acquisition is being performed. For example, the number of milspecs for the Air Force's SBIR (Space Based InfraRed) system has been slashed from 150 to 2; the Navy's SLAM missile decreased from 104 to 6; and the Army's BCIS Phase I antifracticide digital system dropped from 467 to 194.[8]

However, it remains to be seen what savings are to be found from the Federal Acquisition Streamlining Act of 1994. Estimates of 50 percent savings are envisioned by some. If this is true, and if defense readiness is not compromised by the changes, then indeed the commercial marketplace's influence cannot be denied. But the fair skies painted by some when exploiting the commercial marketplace should not construed as being universally accepted or even desirable. The flow of defense conversion is turbulent, and care must be taken when sailing into uncharted waters.

## THE FAILURE OF THE COMMERCIAL MARKETPLACE

The commercial marketplace exists to make a profit[9]—not to be "nice," or to support a lofty, ethereal goal, but for one reason alone: to make a profit. The bottom line of the commercial marketplace is "what is the ROI?"—the return on investment.[10] When a potential for profit is discovered, the resources of the marketplace are thrown on maximizing the ROI. This is precisely why the Department of Defense should exploit the commercial marketplace: there is no better way to leverage the most efficient economic system in the world than by taking advantage of technology spinoffs for "free."

But note that the operative word is *exploit*. For with the alluring promise of a "free" infrastructure, DOD must learn yet a deeper lesson—that the commercial marketplace won't do something for nothing. This is illustrated in the boxed example.

## THE FAILURE TO EFFECTIVELY EXPLOIT DEFENSE LABS

Innovative ideas—concepts that may be destined to ignite a Revolution in Military Affairs (RMA)—provide the winning edge in a future war. In a perfect world, the best solution for a warfighter is to instantly deploy weapons that cause an RMA. These weapons would overwhelm an adversary by striking the enemy quickly, without warning, and with unparalleled force.

Truly revolutionary weapons would forever change the face of warfare. New weapons result in new strategies, new goals, and even a new definition of what it means to "win" a war: Is bombing a command and control center considered more or less of a success than incapacitating an adversary's entire financial system? Such is the power of raw, new ideas that can mature into an RMA. And it is in our nation's best interest to bring them rapidly to the field and into the hands of the warfighter.

"There Ain't No Such Thing As A Free Lunch"

*The Los Alamos National Laboratory used to employ (along with its sister laboratory, Lawrence Livermore) the world's largest cadre of supercomputer operating system programmers. With the introduction of a new supercomputer, the programmers would spend hundreds of man-years furiously refining an operating system so that the rest of the laboratory could perform its flagship work: designing nuclear weapons. The labs developed their own systems and would grudging help each other if problems were encountered. At the time, Los Alamos and Livermore were forced to retain these stable of programmers, as the service was available no where else in the world. And the system worked. However, in the 1980s, once supercomputers began to be used in the commercial marketplace—by oil companies, aircraft manufacturers, automobile designers, and even universities—then a cottage industry quickly grew up to churn out standardized operating systems efficiently. This was an industry devoted to making commercial operating systems and was not simply an aside to some other endeavor, such as making nuclear weapons, for example.*

*Supercomputer manufacturers slowly turned from supporting the Los Alamos and Livermore systems as their marketplace grew in other directions. It took a while, but the labs made a conscious decision to exploit the marketplace and use these "commercial standards." The new standards were more efficient, better maintained, and more responsive than anything that the labs could field on their own. So the labs were able to divert their own people from writing operating systems to performing their unique mission of designing nuclear weapons.*

*This is but one example of how market forces can drive a product. It makes the product more efficient, more cost effective, and even more useful. And it is for these reasons that the commercial marketplace must continuously be scrutinized and combed for where advances might be leveraged. The lesson learned in this example is that the defense research establishment must embrace technology developed outside of its own realm to reduce cost and perform its mission. But the most important lesson in this anecdote is not that the government should exploit the commercial marketplace, but rather that the commercial marketplace did not fill the vacuum without a motive (the "free lunch"). The commercial marketplace needed to assess a clear need that showed the promise of growth, and one of a positive return on investment. There had to be a profit, more than what the national laboratories could give, before the commercial marketplace responded.*

*Industry will not step up to fill a void unless there is a sufficient profit.* Therefore, defense technology needs cannot always be filled by exploiting the commercial marketplace. *There are exceptions to this rule, as is always the case where IR&D (Internal R&D) funding may be used. But it is getting harder for companies to justify spending large amounts of their own money to invest in a high-risk area that may not show a profit and which the government may not later fund. It is because of this void in defense-unique, non-profitable technology that DOD must not treat the commercial marketplace as a panacea. And because*

However, as discussed in chapter 2, these creative ideas may percolate in a research laboratory for decades until they are brought to fruition. Further, it is seldom the *invention* of the weapon that causes an RMA; rather, it is the consequence of the *infrastructure*, or the strategy and application of using a new weapon. Recall that it was the infrastructure associated with ballistic missile technology, and not the invention of the atomic bomb, that caused an RMA. Thus, a laboratory that efficiently transitions ideas out of a basic research stage to an application is key to having a successful R&D program. And that should be the strength of maintaining a defense laboratory system.

## DEFENSE LABORATORIES

In 1980, Director of Defense Research and Engineering (DDR&E) Dr. William J. Perry emphasized that the government must control its research and development effort. Because the growing complexity of defense systems requires understanding the technical issues, in-house technical organizations provide a means to ensure competent leadership. Dr. Perry's goal was to "take maximum advantage of all government, industry, and academic sources to develop the best possible systems and equipment to support our Defense needs. Our in-house technical organizations perform a special role which can be best described as providing the leadership in carrying out these programs."[11]

The DOD laboratory system has evolved over the past two centuries to exploit science and technology fundamental to the unique needs of the services. S&T has been assimilated throughout the services to address operational needs, and DOD labs have accomplished this by successfully communicating with users, collaborating with outside agencies, building (and applying) the technical base, and by developing technically experienced leaders.[12]

It is the White House's National Policy to use the DOD laboratories to:

- Identify the connections between warfighters' needs and technological opportunity;
- Rapidly respond with high-quality technical solutions to warfighters' needs as they develop; and,
- Provide continuity and direct support to acquisition commands—the Program Executive Officers and program managers—through technical expertise, contract management, work force training, and staff support.[13]

One way the laboratories accomplish this is by sending technical people out to the field and by rotating warfighters into the R&D structure to provide cross-feed into the development process. This provides a means to keep in close touch with "operators," and to solve problems encountered by the warfighter. This is absolutely crucial as the whole purpose of having an R&D structure is to ultimately support the warfighter.

However, in recent years DOD labs have been criticized for being unresponsive to warfighter needs. One of the major reasons for this unresponsiveness was a lack of efficiency. In a 1992 S&T strategy discussion between DDR&E and the military services, the issue of using alternate management options was presented as a way to enhance laboratory efficiency. The services identified three "limiting factors" that determined if a laboratory was inefficient: 1) an unclear corporate purpose; 2) unproductive people; and, 3) unrecognized value.[14] Thus, one strategy for increasing laboratory efficiency is to get rid of the limiting factors.

## CRITICAL SUCCESS FACTORS

**The Need for a Clear Corporate Purpose.** Conducting research can be an enthralling, satisfying experience for scientists and engineers. Often, an individual immersed in an intricate aspect of research literally can spend years following a thread through previously uncharted ground. When this happens, too often the connection between the initial purpose of research and the final goal is lost. The individual may get caught up in the details, and the relationship between research and purpose becomes murky.

In the same manner, a laboratory's "macroscopic" research thrust may become unfocused, just as an individual researcher may become unfocused in a "microscopic" sense. *Unless a clear sense of mission is understood throughout the organization, a laboratory may lose its sense of corporate purpose.* The mission and end-product of a laboratory must be clearly defined at every level. The individual researchers must know where they fit in and how the laboratory accomplishes its S&T goals. For example, NASA's Jet Propulsion lab has a clearly defined corporate purpose of excellence in satellites and space technology; the Department of Commerce's National Institute for Standards and Technology (NIST) is a center of excellence for technology, measurements, and standards for industry. In the same manner, the DOD laboratories must continuously reinforce their S&T purpose as being preeminent for land, naval, air and space forces.

**The Need for Productive People**. World-class research is accomplished by world-class people, and excellence in research depends upon the quality of people. Facilities are also important for conducting world-class research—laboratories, material, supplies and equipment—but unless the right people are involved, the best facilities in the world will not enable second-rate researchers to perform as well as first-rate people.

Research supporting U.S. national security goals demands the very best talent. The stakes are high, and the science is complex, so the nation must employ only the very best people in advancing national security S&T. As discussed in an earlier chapter, some of this talent is tapped by exploiting university, commercial and industry researchers. However, defense laboratories must also attract and retain world-class talent. To compete with nondefense laboratories for top-quality researchers, defense labs must go outside the normal civil service rules for hiring (and firing) people.

A recent study concluded that turnover of scientific talent in a research laboratory is a key indicator of success.[15] A high-quality research lab typically turns over 6 percent of its personnel a year. A large fraction of this 6 percent comes from the low performers—those who are not contributing to the lab's corporate purpose and either quit or are fired for their poor performance. A smaller fraction of the 6 percent turnover comes from the high performers, those who contribute greatly to the lab's corporate purpose, but who leave the lab to pursue loftier goals. The vacuum at the top is filled by attracting new talent, identifying and promoting in-house talent, and investing in future prospects.

Management must continuously remain proactive in identifying and moving out low performers, or else the quality of research suffers. Because of restrictive government service rules, DOD labs may evolve from the 6 percent turnover model found in quality institutions to one that allows low performers to remain. This means that the percentage of top performers will decrease. A fraction of these top performers will continue to leave DOD labs—opportunity abounds for top talent—but unless a mechanism is present to deal with *low* performers in the government service system, the number of low performers will increase and mediocrity will set in.

The following chart illustrates this phenomenon, assuming a 6 percent turnover rate concentrated on the bottom performers (left) and on the top performers (right). Note that the percentage of top performers increases if the 6 percent turnover is concentrated on the bottom performers.

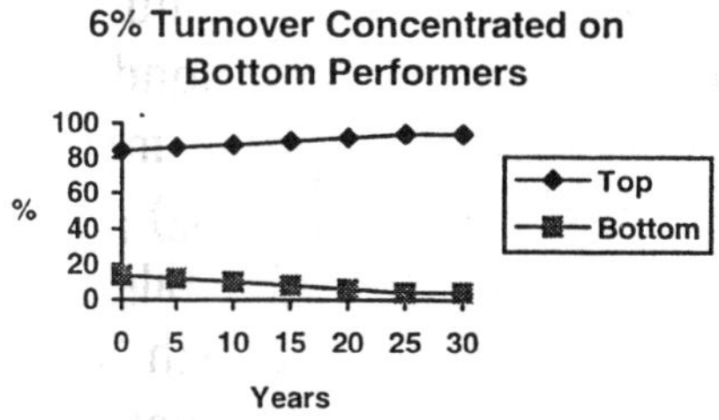


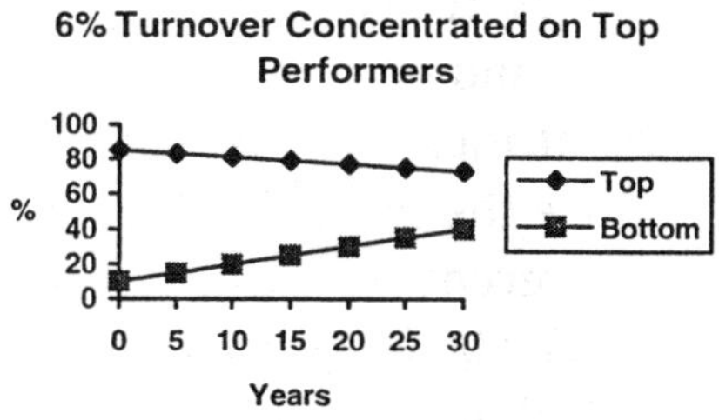


The percentage of top performers are increasing on the left, when the 6 percent turnover rate is concentrated on the bottom performers (i.e. every year 6 percent of the bottom performers leave the organization). When the 6 percent turnover rate is concentrated on the *top* performers (right), the percentage of bottom performers increases and the organization becomes mediocre.
Source: Air Force Chief Scientist

Efforts such as the National Performance Review (NPR) are currently underway to alleviate these problems inherent in the civil service system. One NPR result was the passage of legislation designed to target personnel issues by designating organizations as "reinvention laboratories." NPR recommendations also include allowing DOD labs to conduct their own recruiting, dramatically simplifying the classification system, design their own performance management system, and reduce the time to terminate low-performing employees. Other efforts pioneered at the China Lake Naval Weapons Center and NIST include establishing career paths in competitive areas, establishing pay bands, allowing for the use of local classification standards and for classification to be performed by managers, paying for performance, having flexible entry salaries, and establishing recruiting salaries.[16] Additional personnel initiatives that should be instituted are currently used at

NIST, but not yet in the DOD lab structure, including total compensation comparability, agency-based staffing, local direct hiring for high grades, retention allowances, extended probation, modified qualification standards, new occupational series, paid advertising and recruiting services, linkage of promotion to performance, supervisory pay differential, cash payouts, sabbaticals, and evaluation by contractor.[17]

In addition, management techniques such as putting service agencies under laboratory director control (such as procurement, contracting, civilian hire/classification, and other support services) are absolutely essential. Laboratory Directors need responsive support. When these services are "matrixed out" from other line agencies—no matter how efficient it may look on paper—matrixed functions simply will not be as responsive as when the service agency is directly accountable to a laboratory director.

**The Need for Recognized Value.** The final critical success factor is the need for DOD laboratories to have a clearly recognized value throughout their service, especially at the highest levels of defense R&D. This value is diminished when the lab has *multiple distinct reporting lines*; that is, when the lab is *not* responsible to a single chain of command but to several different authorities. Some real-world examples include being simultaneously responsible to:

- A higher *funding* authority (that distributes money)
- A higher *command* authority (that gives policy guidance)
- A higher *technical requirements* authority (that gives technical guidance as to what kind and what mix of research needs to be accomplished—basic, applied, or advanced development)
- A higher *personnel* authority (that determines what type of personnel is needed at the lab—e.g. how many physicists, engineers, etc.).

Although multiple reporting lines have the perceived advantage of making the labs responsive, in reality the lab is forced to fight against a constant barrage of outside demands, some of which are not in line with the lab's corporate purpose.

Therefore, just as internally the Laboratory Director must be able to exercise a direct line of control, *externally the laboratory must have a clear, direct reporting line to a relatively high-level of authority*. For example, the following table shows the reporting level and technical reputation of several highly successful non-DOD laboratories:

**Table 4.** ***Importance of Reporting Level (Recognized Value) as a Success Factor***

| *Lab* | *Reporting Level* | *Technical Reputation* |
|---|---|---|
| Lincoln Labs | MIT Provost , DDR&E | Preeminent in field |
| Jet Propulsion Lab | CalTech President, NASA Administrator | Preeminent NASA lab |
| Los Alamos National Lab | UC President, Assistant Secretary of Energy | Perceived excellence in basic and applied science |
| NIST | UnderSecretary of Commerce | Perceived excellence in measurements, standards |

All DOD laboratories are currently constrained by Office of Personnel Management regulations, in effect, a "separate" reporting line that determines laboratory demographics. Commandwise, the Naval Research Lab (NRL) reports directly to the Assistant Secretary of the Navy through the Chief of Naval Research—a clear, single reporting line—and has a national reputation of excellence. The Army's Research Lab (ARL) reports directly to the

Assistant Secretary of the Army. Although it does not have the national reputation as NRL, it is improving. The four Air Force "superlabs" (Armstrong, Phillips, Rome, and Wright) report to their respective Air Force Material Command (AFMC) center commanders for personnel concerns and report to the AFMC director of S&T and the AF deputy assistant secretary (acquisition) for program direction and budget; the AF labs have made nationally significant contributions (some for which they have not received credit) but have less of an in-house national reputation than NRL, focusing instead on leading industry teams to develop technology for Air Force needs.[18]

DOD laboratories must continue to ensure their value is recognized by both "owners" and "customers." *Executive leadership can help and can greatly add to the lab's national reputation by ensuring the labs have a clear, direct line of authority (both externally and internally), and by having the labs report to the highest level possible.* Only then can a laboratory be free to full-fill its true corporate purpose: supporting the warfighter.

## THE FAILURE TO USE COMMON SENSE WHEN CONDUCTING R&D

### IMPROVING R&D EFFECTIVENESS

Although the evolutionary process of going from a creative idea to the application of a weapon system has been laid out in a sterilized format (in the form of 6.1 through 6.5 programs), in reality the procedure is much more complex. The process is not seamless. There is no one person who oversees a weapon system from idea to Initial Operating Capability (IOC), or in the vernacular, from "criteria to killer." Having continual, seamless oversight is one way to field a weapon system and ensure that every development phase is adequately mature, but because of the enormity of the problem, it is impossible in practice to do such a task.

Although much has been said about how the various elements of Program 6 overlap, there exist distinct "cultures" or "tribes" that rule over each stage of the development process—basic research, exploratory development, advanced development, acquisition, and procurement. Although there are no hard and fast rules about the qualifications of the people who may participate at each stage, for the most part the cultures are "stovepiped" in the sense that very little crossflow exists among the cultures.

For example, in the past, scientists—physicists, chemists, biologists, and applied scientists (who may be defined as engineers performing at the basic research level)—typically dominated the 6.1 (basic research) and 6.2 (exploratory development) research phase. Engineers, particularly mechanical and electrical engineers, populated the 6.2, 6.3 and some of the 6.4 phase. Acquisition specialists with some technical background are involved starting at the 6.3 level. The "operators," or warfighters, generally do not get involved until the very last phase, when a technology demonstration of the weapon system is being performed.

Such segregation may result in a perception of "hobby shopping" or "playing in a sand box" at the 6.1 or 6.2 level, because communication with the end-user—the warfighter, usually the one with the funds—is nonexistent. The S&T process may be entirely misrepresented to the warfighter because the huge investment of laboratory infrastructure may not be commensurate with the final product that the warfighter uses. This is realized at the national level in a statement from the National Science and Technology Council:

> A balance of investments is needed at every phase of development to ensure that basic research results are exploited for military applications in a timely manner through technology demonstrations and transition. Further, since there are many performers, we must carefully manage our investments to make sure that we capitalize on all their strengths.[19]

A problem exists if a direct link cannot be shown from the creative echelons of a weapon system development phase (6.1 and 6.2) to the final product. Tis is not a new problem; Winston Churchill also recognized the essence of the conflict that existed between the basic researcher and the warfighter:

> A hiatus exists between the inventor who knows what they could invent, if they only knew what was wanted, and the soldiers who know, or ought to know, what they want and would ask for it if they only knew how much science could do for them. You have never really bridged that gap yet.[20]

Some concrete metrics have been used in the past to quantify the effectiveness of laboratories, including the number of PhDs at a lab, the budget, the total number of people, the unique facilities, the number and quality of universities near to the lab, the number of CRDAs (Cooperative Research and Development Agreements with industry), tallying the number of peer-reviewed papers, the number of citations the papers have received by other papers, the number of patents, and the number of awards given to the laboratory.[21]

The commercial world uses these metrics, plus another, to test the ultimate effectiveness of a lab: "The creation of new businesses or competitive advantage for *quantum business growth* [that] justifies the existence of a corporate R&D center, especially in these days of close scrutiny from corporate headquarters."[22] The situation is no different in the Department of Defense. But instead of creating new commercial businesses, the DOD laboratories should be in the business of creating new breakthroughs for advanced weapons—weapons that will produce the next Revolution in Military Affairs. *Thus the effectiveness for a defense laboratory should be measured by the success of the technological weapons in the field, success measured by winning the war.*

Certainly DOD laboratories must contribute to other, less lofty goals than continually striving for creating an RMA. DOD laboratories have a place in identifying and exploiting those areas that the commercial market has ignored, and by optimizing and refining current weapon systems while advancing the state-of-the-art. This is akin to having a mix of ball players on a team, some who are solid base hitters, while others go for the home run; both are needed to win a game.

## THE SKUNK WORKS APPROACH

The defense research establishment is controlled by legislative and regulatory constraints, some of which even seem to work against the purpose of having a laboratory system. There is much to be learned from studying commercial research organizations, and applying their ways of conducting research. One of the most successful advanced technology establishments in the world for rapidly evolving new concepts into engineering systems is the Lockheed "Skunk Works," located in Palmdale, CA. Skunk Works has come to be associated with an aggressive management style, characterized by the extremely efficient use of time by a small, dedicated group of highly skilled and creative scientists and engineers.

Skunk Works' efficiency is accomplished by having lucid, clearcut goals, which are derived from several focused, immutable strategies:[23]

- *Planned* contacts—in the sense of being regularly scheduled and limited by "contract" with upper management—with market research teams and potential customers
- An optimized, custom-designed workplace located near—but physically separate from—company facilities
- A permeating sense of urgency, attention to detail, and quality

- A complete absence of "red tape"
- A short managerial chain-of-command, which not only allows the team to remain focused and decline urgent but unrelated projects, but also "flies top cover"—provides support at the application phase of the innovation process.

One criticism of Skunk Works is that it can thrive only in industries "subsidized" by the government, where high-risk endeavors are actually "only" pushing the state-of-the-art. But the Skunk Works philosophy of business is not limited to the aircraft industry or to pushing the state of the art. Several highly successful Skunk Works have been established in various major corporations, including in such unlikely places as automobile manufacturers and the Ballistic Missile Defense Organization's *Clementine* small-satellite group.[24]

The Lockheed Skunk Works lives and dies by 14 operating rules, written over 50 years ago by Skunk Works founder Kelly Johnson, that govern the way they do business:[25]

1. The Skunk Works program manager must be delegated practically complete control of his/her program in all aspects and should have authority to make quick decisions regarding technical, financial, or operational matters.
2. Strong but *small* project offices must be provided by the military and the industry.
3. The number of people having any connection with the project must be restricted in almost a vicious manner. Use a small number of good people.
4. Very simple drawings with great flexibility for making changes must be provided in order to make schedule recovery in the face of failures.
5. There must be a minimum number of reports required, but important work must be recorded thoroughly.

6. There must be a monthly cost review covering not only what has been spent and committed, but also projected costs to the conclusion of the program. Don't have the books 90 days late and don't surprise the customer with sudden overruns.

7. The contractor must be delegated and must assume more than normal responsibility to get good vendor bids for subcontract on the project. Commercial bid procedures are often better than military ones.

8. The inspection system as currently used by the Skunk Works, which has been approved by both the Air Force and the Navy, meets the intent if existing military requirements and should be used on new projects. Push basic inspection responsibility back to the subcontractors and vendors. Don't duplicate so much inspection.

9. The contractor must be delegated the authority to test his final product in flight. He can and must test it in the initial stages.

10. The specifications applying to the hardware must be agreed to in advance of contracting.

11. Funding a program must be timely so that the contractor doesn't have to keep running to the bank to support government projects.

12. There must be absolute trust between the military project organization and the contractor with very close cooperation and liaison on a day-to-day basis. This cuts down misunderstanding and correspondence to an absolute minimum.

13. Access by outsiders to the project and its personnel must be strictly controlled.

14. Because only a few people will be used in engineering and most other areas, ways must be provided

to reward good performance by pay and not based on the number of personnel supervised.

The government has proved that it can use the Skunk Works philosophy in day-to-day operations. The success of the Ballistic Missile Defense Organization's *Clementine* program showed that a small team could send a satellite to the Moon and do so with a minimum of cost and oversight. In the same manner, DOD labs could become even more efficient if a Skunk Works mentality permeated their culture.

## SUMMARY

This chapter has identified three "kinks" in the national S&T security strategy that may overwhelm even the best laid plans to employ a robust R&D program:

- Failure of the commercial marketplace to step up and fill a void left by the dismantled defense R&D establishment
- Failure to exploit the defense labs effectively
- General failure to use common sense when conducting R&D.

These observations are by no means all inclusive; they are merely the most apparent after analyzing the turbulent changes going through defense R&D. Along with identifying some "kinks" in the strategy, several suggestions were presented to smooth the kinks, including realizing that 2) Defense technology needs cannot always be filled by exploiting the commercial marketplace; 2) three critical factors determine the success of a lab: a clear corporate purpose, productive people, and recognized value; and 3) the effectiveness for a defense laboratory should be measured by the success of the technological weapons in the field, success measured by winning the war.

The next chapter discusses specific recommendations for policy makers and lab directors if they are really serious about optimizing S&T.

## NOTES

1. Deputy Director of Defense Research and Engineering, *Defense Science and Technology (S&T) Strategy*, unpublished briefing given to Congressman Dicks and his personal staff, December, 1995.

2. Paul Kennedy, *Preparing for the Twenty-First Century* (New York: Random House, 1993), 293.

3. A classic argument is that the United States has diverted R&D funds from enhancing the commercial sector to the military—moneys that the military is using to build weapons and maintain its infrastructure. However, this presupposes that the funding would have been redirected from the military to the commercial sector, and specifically to commercial R&D. Paul Kennedy covers this point in ibid., 294.

4. Kenneth L. Adelman and Norman R. Augustine, *The Defense Revolution: Intelligent Downsizing of America's Military* (San Francisco, CA: Institute for Contemporary Studies Press, 1990), 110.

5. Jacques S. Gansler, *Defense Conversion: Transforming the Arsenal of Democracy* (Cambridge, MA: MIT Press, 1995), 177.

6. Director of Defense Research and Engineering, *Defense Science and Technology Strategy* (Washington DC: Department of Defense, September 1994), 29.

7. Executive Office of the President, Office of Science and Technology Policy, *National Security Science and Technology Strategy* (Washington, DC: U.S. Government Printing Office, 1995), 19.

8. Ibid., 19.

9. William J. Baumol and Alan S. Blinder, *Economics: Principles and Policy*, 6th edition (New York: Harcourt Brace & Company, 1994), 108.

10. This fact is sometimes lost on the altruistic. The commercial marketplace does not exist to cure social ills, to provide for the common defense of our country, or to pave the way for a better future. Some people in the marketplace have these goals, and the world is a far better place because of them—numerous organizations exist that are openly nonprofit and exist for the common good. However, unless a profit motive is shown for what a *commercial* corporation does, then the stockholders will not put up with that situation for long.

11. Executive Office of the President, Office of Science and Technology Policy, 17.

12. U.S. Air Force Chief Scientist Briefing to DDR&E: Blue Ribbon Panel on Laboratory Management, 1993.

13. Executive Office of the President, Office of Science and Technology Policy,17.

14. U.S. Air Force Chief Scientist Briefing.

15. Ibid.

16. Ibid.

17. Ibid.

18. Ibid.

19. Executive Office of the President, Office of Science and Technology Policy, 16.

20. Winston S. Churchill, *The Great War*, vol. 4 (London: The Home Library edition), Appendix 8, War Memorandum, "Mechanical Power in the Offensive," 9 November 1916, quoted in Michael I. Handel, *War, Strategy and Intelligence*, 131.

21. The Naval Research Laboratory (NRL) proudly—with justification—heralds the presence of the "Only Noble Prize Winner at a DOD laboratory."

22. Ted M. Foster, "Making R&D More Effective at Westinghouse," Research Technology Management, January-February 1996, 31.

23. Arthur W. Single and William M. Spurgeon, "Creating and Commercializing Innovation Inside a Skunk Works," *Research Technology Management*, January-February 1996, 39.

24. Ibid., 38-41.

25. Ben Rich and Leo Janos, *Skunk Works: A Personal Memoir of My Years at Lockheed* (New York: Little, Brown and Company, 1994), 51-52.

# 6. RECOMMENDATIONS

> With the downsizing of defense must come more, not less, risk taking on revolutionary new weapon concepts. . . . Qualitative advantage grows more important as quantity shrinks and more difficult as technology becomes more diffused throughout the world.
>
> Kenneth L. Adelman and Norman R. Augustine
> *The Defense Revolution: Intelligent Downsizing of America's Military*

> The key to successful exploratory development, and to R&D as a whole, is an environment where technically sound managers with entrepreneurial instincts have flexibility to act and the ability to delegate technical decisions to those who are executing the programs. The managers must be able to set goals, select projects, monitor execution, pick winners, and eliminate losers without making the detailed decisions within the projects they supervise.
>
> Gregory H. Canavan and John F. Lilley
> *Defense Technology*

Throughout the years there have been many studies detailing not only what is wrong with the present DOD R&D structure but what should be done to rectify the apparent problems. The studies include Defense Science Board reports, National Research Council studies, the Air Force's *New World Vistas*, and the Army's *Star 21*, among others.[1]

From this effervescent and sometimes self-conflicting data base of quick-fixes, "silver bullets," and sage advice, a few of the recommendations that best address the problems outlined in chapter 4 have been culled. This is certainly not an exhaustive list, as there is a wealth of advice in the literature for improving the defense science and technology infrastructure. In consideration of the frustrations these recommendations are targeting, the major failings of defense R&D are highlighted. The literature also contains some solid advice for "filling in the holes." This is accomplished by tackling those issues not addressed by the administration's policy of DOD R&D

Chapter 4 highlighted three major failures of national security R&D strategy:

- Failure of the commercial marketplace to step up and completely fill the void left by the dismantled defense R&D establishment
- Failure to effectively exploit the present defense laboratories
- General failure to demand the use of common sense when conducting R&D.

The solutions to *some* of these problems transcend any recommendations that this book or anyone else can give: they demand attention from the highest leadership in the form of specific policy guidance or legislation. (For example, overturning some Office of Personnel Management regulations requires congressional intervention.) However, there *are* some problems that can be solved by Laboratory Directors or by individuals at the Assistant Secretary-level—the target audience for this book. It is therefore in the spirit of improving the overall DOD R&D infrastructure—by using common sense and judiciously implementing the following recommendations—that this advice is given.

This chapter lists various "threats" to the DOD R&D establishment, with a particular emphasis on the threats to performing S&T, specifically to basic research. The following addresses the shortcomings of defense R&D strategy as raised in this book and urges policy makers and directors or commanders of laboratories to beware their insidious presence.

## RELYING TOO MUCH ON THE COMMERCIAL MARKETPLACE

The failure of the commercial marketplace to fill a void left by the dismantled DOD R&D infrastructure has already been identified as a flaw in the overall defense S&T strategy. The specific *threat* is that DOD may be forced to rely on the marketplace, and may not find it always responsive, particularly in times of crisis. Thus, the lesson for policy makers and lab directors is that *defense technology needs cannot always be filled by exploiting the commercial marketplace.* In other words, industry will not step up to fill a void unless there is a potential for profit.

There are exceptions to this rule, as is always the case where IR&D (Internal R&D) funding may be used. But it is getting harder for companies to justify spending large amounts of their own money to invest in a high-risk area that may not show a profit and in which the government may not later fund. The government cannot expect companies to commit their own money when there is no assurance to recoup losses. Thus, a void will exist in certain technology areas.

It is because of this void in defense-unique technology that *DOD must not treat the commercial marketplace as a panacea.* One way is for the government to throw money at the problem—give the contractor the unrestricted right to produce whatever defense technology they decide is best for the military. Another, better way is for defense laboratories to partner with contractor, industry, and university alike. In effect, this supplements core competencies found in defense labs with expertise drawn from industry and universities. By partnering, defense labs can pull in experts when they're needed, keep on top of ongoing research, and still direct the final outcome without having to stand back as an outside (and sometimes antagonistic) observer.

But the *best* way to ensure that services acquire the right technology for the warfighter—be it technology grounded in the commercial marketplace or uniquely defense—is by training and keeping "knowledgeable warriors," the topic of the next section.

## NOT HAVING KNOWLEDGEABLE WARRIORS

Some have argued that DOD should simply "sow the seeds" for industry by funding the hell out of research, stepping back, and waiting for the magic to happen: watch technological solutions appear like a rescuing knight from the mists of industry. However, arguments presented in chapter 3 show the necessity for DOD to have "smart buyers" and competent technical liaisons, and this is possible only if DOD keeps abreast in the advances in science and technology by having a cadre of competent scientific personnel—"knowledgeable warriors," competent in S&T, able to pierce through the issues, and serve as an honest broker to insure technological progress is oriented toward the warfighter. This means training warriors in science and technology, and putting them in charge of defense-related S&T activities.

Another way to look at this is that the services would not hesitate to put a doctor in charge of medical care, or a lawyer in charge of resolving legal issues, or a pilot in charge of flying an aircraft. Ensuring that the best S&T is tapped for national security demands no less of a scientific professional overseeing S&T concerns. *Not* having a cadre of scientifically competent personnel in the services responsible for overseeing S&T is akin to putting someone with no acquisition experience in charge of a major acquisition program, and DOD must apply the same analogy to S&T.

Some argue that, because the mission of the military is to fight and win wars, this cadre of scientific personnel should consist only of civilians—leave the warfighting to the warriors and allow this civilian cadre to generate new weapons of war. There is a good point to this argument, that the military should focus on its core competency—serving at the "pointy end of the stick." However, the world has changed from the times when the warfighter could just be handed a piece of equipment and be unthinkingly ordered into battle. Science and technology are inextricably tied to the weapons of war, and a sophisticated weapon system demands that no breaks occur from its birth to its use. The only way to ensure that this continuity exists is for a dedicated cadre of scientifically smart warriors—those who are closest to the warfighters carrying the equipment into battle—to maintain the continuity.

These are the people who can transition a discovery in basic research—a creative concept—to a weapon that satisfies a warfighting requirement. Carrying the fledgling weapon across the "seams" through the sophisticated stages of the development process will ensure continuity. The technically competent warriors need to be scientifically "smart" to recognize the future utility (or futility, as illustrated in the Navy's A-2 debacle) of a discovery, and pull the concept up to become a militarily useful weapon. They must have the authority to keep the weapon focused on it's ultimate use: to support the warfighter. They must ensure that it doesn't start to be burdened with additional requirements, usually put on the weapon by well-meaning but totally clueless bureaucrats. Further, they must follow the weapon from the scientific bench to the hands of the warfighter, providing a seamless transition so that the system doesn't get held up by unnecessary delays. The only way to accomplish this important transition process is to have competent personnel that have credibility both with the *science* and with the *warriors*. For only then will the ultimate goal of defense S&T be accomplished—supporting the warfighter's needs.

## INSUFFICIENT FUNDING

Although much is being said about preserving the S&T base—and for the last 10 years the S&T budget has remained relatively stable—the fact is that the defense basic research and exploratory research (Program 6.1 and 6.2) budget as a percentage of total federal RDT&E has decreased by over 31 percent in the past 30 years.[2] This is in spite of the administration's pronouncement that "New technologies have dramatically enhanced our ability to both prepare for and execute military actions."[3]

What is occurring is simple to grasp: within S&T, more advanced development programs are being funded at the expense of basic research and exploratory development. Advanced development programs are more expensive to perform than basic research and exploratory development programs, and a larger fraction of money is being diverted into proving concepts that have a military application. Although advanced development programs are indeed the vehicles that will transition the discoveries made in basic science to the warfighter, *they are in danger of choking off creative input, and will therefore only produce evolutionary gains, and not true revolutionary or "leapfrogging" advances.*

It cannot be overemphasized that funding for the creative, or "invention" phase—basic research and exploratory development—has *decreased* as a percentage of defense RDT&E. Therefore, if we are to expect to achieve the same degree of technological superiority in the future as we have in the past, then that percentage must also increase to the historical levels of the past. This does not mean that advanced development should take a backseat to basic research. Rather, in order to ensure a continued robust advanced development effort, the final product (a weapon to the warrior) is only as good as the initial conditions (the basic research)—and production can always be stretched out. So if there is a "zero sum game" to be played with funding, then the decrease should come from the *production* phase and not from the initial phase. Produce when we need it, develop and refine until then, but ensure only the *best* creative ideas are pushed forward to develop. Therefore, basic and applied research must be *strengthened* (and not at the expense of advanced development) so that quality input may be given to advanced development efforts.

## CENTRALIZATION

The more DOD centralizes R&D, the more it will control programs from the top. More control from the top not only means the necessity for more layers of management bureaucracy, but it also means *less* control for individual program managers. This results in a loss of authority for program managers, and a diffusion of responsibility for the program.

The perceived need for centralized control evolves from three reasons:[4]

- A continuing increase in management oversight
- A greater flow of information to the top in response to the increase in management oversight (i.e. the phenomenon of nanomanagement)
- A replacement of normal healthy tension between headquarters and laboratories with an atmosphere of blind compliance.

An increase in bureaucracy occurs naturally for every organization unless specific steps are taken to snarl its growth; this cross-cutting truth is illustrated in every organization fromDOD to NASA.[5] The problem was astutely recognized in 1983 by the White House Science Council's report on federal laboratories:

> Excessively detailed direction of laboratory R&D activities from agency headquarters . . . has seriously impaired R&D performance in some laboratories. Numerous detailed external directions are given as to how work should be done, while at the same time, the overall missions and goals of the laboratories are inadequately defined.[6]

The solution is to consciously decentralize.[7] This is best illustrated by a recommendation from the *Synthesis Group*, a vice-presidential commission led by Astronaut Lieutenant General (USAF, Ret) Thomas P. Stafford, when the group outlined architectures for controlling multiple government agencies that will work on a national program:

> Program management must address such functions as systems architecture planning and requirements control. The responsibilities at the headquarters level are to identify key program requirements and review the implementation of programs, but not execute the activities. Such detailed design and development is best accomplished at the field levels.
>
> Consolidating federal resources, where appropriate, and guarding against the effects of institutional aging and development of bureaucratic culture[8], will be key in deriving the plan.[9]

## SPECIFIC RECOMMENDATIONS

To conclude their landmark text *The Defense Revolution: Intelligent Downsizing of America's Military*, Adelman and Augustine include several "Memos to Policy Makers." The memos allow the authors the ability to play "if I were King," and give advice to government policy makers in critical positions, including the U.S. President, the President of the Soviet Union (reflecting the text's 1990 publication date, but just as well suited for today's President of Russia), the Secretary of State, Members of Congress, the Secretary of Defense, the defense industry, and to the U.S. Under Secretary of Defense for Acquisition—not only the "third ranking" member of DOD, but the individual responsible for defense R&D. The memos are frank and show a promarket bias; but this is only one approach for refining the Department of Defense R&D infrastructure. Other writers, such as Canavan and Lilley, posit another approach.[10]

The following is a short compilation of "nuggets," advice culled from government, academic and commercial executives who have worked in the DOD R&D environment for years. They provide an anchor for solving the specific threats listed above.

- Assign first priority to research and early exploratory development that promise major breakthroughs in capability or lead to entirely new conceptual approaches for performing important mission.[11]
- Do not rely *only* on university and industrial ("commercial") basic research when reducing the government's in-house overhead structure. Recognize that government laboratories play the *key* role in performing research where there is no commercial market.[12]
- Once the key role of government labs is recognized, do not either take over or absorb the military departments research, development, and production activities into a centralized DOD organization.[13]
- Seek to apply new technology not only in the traditional manner—to increase operational capability—but also to decrease cost.[14]
- Make it demanding to initiate new full-scale development programs, insisting on truly significant potential military benefit from all programs that are approved.[15]
- Draw as heavily as possible on the country's commercial (non-defense) production base by standardizing commercial

components wherever practical, and designing to use our commercial manufacturing base, especially in times of emergency.[16]

- Increase substantially the effort to build R&D prototypes. R&D prototypes cost much less and have greater flexibility, all the while preserving critically important design teams.[17] In other words, invest in "X-Programs" that prove concepts.
- Carry no system beyond R&D prototyping (that is, into engineering development) unless there is a firm intent to manufacture and deploy it.[18]
- Pursue "breakthrough systems" based on new concepts and on new technology into engineering development and production—but insist on "fly-before-buy" demonstrations.[19]
- Produce those systems that do in fact receive production go-ahead at the most efficient rate and then shut down the lines, preserving tooling, and in some critical cases stockpiling limited numbers of components that have particularly long lead times.[20]
- Prolong the life of "platforms" (aircraft, ships, tanks) as long as possible by introducing new technology from time to time as system upgrades. This appears to be the only way to keep platforms that must last 40 or more years moderately current with technology that has a half life of 10 or fewer years.[21]
- Seek cooperative programs to buy and sell hardware from and to allied nations to stretch defense dollars and help keep production lines warm where practicable. Be prepared to add our own countermeasure provisioning to these systems. On the other hand, avoid joint development projects—such projects are hard enough with a *single* customer.[22]
- Employ only the amount of concurrency (overlap between consecutive phases of a program) that is needed to maintain an efficient program plan, since time is less critical in the new geostrategic environment.[23]

- Establish an office to address and eliminate impediments to potential wartime reconstitution of a large-scale defense industrial base—mobilization—should it again be required.[24]
- Once programs have passed the new, exceedingly stringent go-ahead gate, fire the first person or organization that meddles with funding, schedule, or requirement—including Congress, if multiyear funding cannot be approved. Tough out the problems and finish things once started.[25]
- Guard against procurement's four horsemen of the apocalypse: turbulence, adversarialism, inadequate personnel, and insufficient margins. [26]

## CONCLUSIONS

The strength of our nation relies on its economic and military might, and underpinning this strength is the health of the American industry, for it is industry that drives business, the economy, and ultimately the quality of our military's weapons of war. Just as science is crucial to the health and conduct of modern industry, science is also the foundation for technology—the fuel that powers the engine of change that advances our strength.

The military victories of the past owe much to the scientific and technological edge our armed forces have assimilated, striving for quality over simple quantity. Much investment has been made to ensure this technological edge, and to a large degree, it is the advances springing from long-range research that have made the difference. These advances have taken years to come to fruition, but ultimately, investments in long-range research have *always* paid off.

The nation has a long and rich history of supporting defense-oriented research and development. For years there has been a tight coupling between scientific advances and sophisticated weaponry, and for years we have seen an ever-increasing march of technological achievements: precision-guided weapons, global positioning satellites, stealth-weapon platforms. However, with a renewed emphasis put on supporting the warrior and paying attention to the ever-present bottom line of cost, it is generally realized that things must change, for we live in a different world, one where we are not at war, and one where the priorities of our nation have changed. But they must *not* change so much as to impair our future warfighting capability by cutting long-range research.

In a year of study of DOD's post-Cold War S&T strategy, great strides have been made to ensure that a well-poised, robust R&D infrastructure exists in DOD. Little by little the obstacles are chipped away and the system is made "better," sometimes not as fast, or as not as much as some people desire, but things are changing. *Times* are changing, and for the best.

Every person interviewed for this book was concerned about the present and future state of defense R&D. This is encouraging. No one is willing to accept the status quo or willing to coast, to go on without trying to change things for the better. Everyone knew that there are problems with "the system," and everyone voiced a different solution.[27]

The problems are not new or unique, but they are not insurmountable. This book has attempted to make sense of the defense R&D infrastructure—a short history behind long-range research, the rationale for it, the structure it has taken, and some recommendations for using common sense for keeping it going. As stated at the beginning of this text, if there is any one thing that our generation can improve to ensure our children's freedom, it is by investing in the future—by sanely supporting long-term research in the basic sciences.

## NOTES

1. Genevieve J. Knezo compiled a wealth of information, detailing numerous studies, committee reports, and publications on re-prioritizing DOD R&D in her exhaustive report *Defense Basic Research Priorities: Funding and*

*Policy Issues* (Washington, DC: Congressional Research Service, Library of Congress, CRS 90-506, October 24, 1990).

2. Ralph Sanders, *Defense Research and Development* (Washington, DC: Industrial College of the Armed Forces, 1968), 18. and Aerospace Industries Association, Inc. *Aerospace Facts and Figures 1995-1996* (AIA Press: Washington DC, 1995), 103.

3. Executive Office of the President, Office of Science and Technology Policy, *National Security Science and Technology Strategy* (WashingtonDC: U.S. Government Printing Office, 1995), ii.

4. Gregory H. Canavan and John F. Lilley, "Managing Public Sector Research and Development: Innovation Versus Responsiveness," in *Defense Technology*, A. Clark IV and J. Lilley, eds. (New York: Praeger, 1989), 253.

5. The number of NASA employees (excluding contractors) grew from a low of 17,077 in 1961 quickly to a high of 33,924 in 1966 to support the highly successful Apollo program. After downsizing in the mid-1970s to around 22,000 employees, the number increased to 24,513 in 1992 and has decreased to 23,000 in 1996 (estimated)—with no corresponding decrease in management structure, although the present Administrator, Dan Goldin, is currently working to decrease headquarters staff by up to 50 percent. The point of these statistics (all from *Aerospace Facts and Figures: 1995-1996*, Washington, DC: Aerospace Industries Association of America, Inc., 1995, 150) is to illustrate that unless a conscious effort is made to continuously keep the bureaucracy in check, it will *always* grow.

6. Canavan and Lilley, 254.

7. The strategy for keeping management focused only at the strategic level, and not on the tactical level (i.e. focusing on "the big picture" rather than "getting down in the weeds") is oft opined in the management lexicon of Total Quality Management as "empowerment." However, the majority of managers who regurgitate this philosophy do so not because they truly believe it, but because it has been dictated from above: "They see but they do not believe." In a talk before the Industrial College of the Armed Forces, The Chairman of the Joint Chiefs of Staff, General Shalikashvili, concentrated on this fact with the illustration that a superior will only "empower" an employee until the superior is called into question about a subordinate's decision; and instead of defending the subordinate and trying to find out the true situation, the superior will more often then not sweep away any pretense of empowerment and resort to more draconian measures of management.

8. A key recommendation discussed among the *Synthesis Group* staff for success is to ensure that for any future program no more than one lawyer be assigned to the program per billion dollars, and that after every seven years the

entire organization be dismantled—corporate memory would best be accomplished by reviewing written records rather than fighting entrenched dinosaurs.

9. Astronaut Lieutenant General (USAF, Ret.) Thomas P. Stafford, The *Synthesis Group* report to the Vice-President: *America at the Threshold: America's Space Exploration Initiative* (Washington, DC: U.S. Government Printing Office, 1991), 85-86.

10. Canavan and Lilley, 253.

11. Kenneth L. Adelman and Norman R. Augustine, *The Defense Revolution: Intelligent Downsizing of America's Military* (San Francisco: Institute for Contemporary Studies Press, 1990), 226-228.

12. This is my departure from Adelman and Augustine's recommendation. The commercial market is not a panacea—it can and must be leveraged. However, for those areas where the commercial marketplace will not (or cannot, because of risk or expense) tread, in-house government laboratories must lead the way.

13. I've modified this slightly from Canavan and Lilley, 244-263.

14. Adelman and Augustine, 226-228.

15. Ibid.

16. Ibid.

17. Ibid.

18. Ibid.

19. Ibid.

20. Ibid.

21. Ibid.

22. Ibid.

23. Ibid.

24. Ibid.

25. Ibid.

26. Ibid.

27. These solutions run from the benign to the radical. For example, one person suggested that the system could *never* be fixed, and therefore should be razed and started over with a clean slate; another insisted that giving Laboratory Commanders (or Directors) the same authority in hiring, firing, and budgetary discretion (freely moving funds among Program 6.1, 6.2 and 6.3 accounts to accomplish program goals and their strategic vision) as their Department of Energy counterparts (at Los Alamos or Lawrence Livermore National Laboratory), would take care of the problem. The truth, as always, lies somewhere in between—or, is at the least a linear combination of the possible solutions.

# BIBLIOGRAPHY


Adelman, Kenneth L., and Norman R. Augustine. *The Defense Revolution: Intelligent Downsizing of America's Military*. San Francisco: Institute for Contemporary Studies Press, 1990.

"Assessing the Potential for Civil-Military Integration: Technologies, Processes, and Practices." Office of Technology Assessment, U.S. Congress (September 30, 1994).

Babbin, Jed L. "Real Defense Conversion Requires Tax Relief for R&D." *Aviation Week and Space Technology* (August 8, 1994): p61.

Becker, Pamela. "Is the TRP the Defense Conversion Cure-All?" *Mechanical Engineering* (October 1994): 32.

Broadhead, John. "Planning a Successful R&D Alliance." *Research Technology Management* (September-October 1995): 2-13.

Canavan, Gregory H., and John F. Lilley. "Managing Public Sector Research and Development: Innovation Versus Responsiveness." In *Defense Technologies*. New York: Praeger, 1985.

Cooper, Robert G. "Developing New Products On Time, In Time." *Research Technology Management* (September-October 1995): 49-57.

Cordtz, Dan. "Bye-Bye, Dr. Strangelove." *The Economist* (January 17, 1995): 32-37.

Cusumano, Michael A., and Richard W. Selby. "How Microsoft Competes." *Research Management Technology* (January-February 1996): 26-37.

"Defense Basic Research." Office of the Director of Defense Research and Engineering. Washington, DC (May 1995).

"Defense Science and Technology Strategy." Director of Defense Research and Engineering. Office of the Secretary of Defense (July 1992).

"Defense Science and Technology Strategy." Director of Defense Research and Engineering, Office of the Secretary of Defense (September 1994).

"Defense Science and Technology (S&T) Strategy." Deputy Director, Defense Research & Engineering, unpublished briefing, 1995.

"Department of Defense In-House Laboratories." Report of the Defense Science Board Task Force, Office of the Director of Defense Research and Engineering. Washington DC (31 October 1966).

Draheim, Kirk, Richard P. Howell, and Albert Shapero. "The Development of A Potential Defense R&D Complex." SRI Project No. IMU-4370. Stanford Research Institute, Menlo Park, CA (July 1966).

Edelheit, Lewis S.. "Renewing the Corporate R&D Laboratory." *Research Technology Management* (November-December 1995): 14-18.

Foerstel, Herbert N. *Secret Science: Federal Control of American Science and Technology*. Westport, CT: Praeger, 1993.

"FY 1994-99 Future Years Defense Plan." Defense Science Board Task Force Report. Office of the Secretary of Defense, Washington DC, May 1993.

Gansler, Jacques S. *Defense Conversion: Transforming the Arsenal of Democracy*. Cambridge, MA: MIT Press, 1995.

Howell, Richard P., and Albert Shapero. "The Structure and Dynamics of Research and Exploratory Development in Defense R&D Industry." SRI Project No. IMU-4370, Stanford Research Institute, Menlo Park, CA, August 1966.

Johnson, Stuart E., and Martin C. Libicki. *Dominant Battlespace Knowledge: The Winning Edge*. Washington, DC: National Defense University Press, 1995.

Kelley, Maryellen R., and Todd A. Watkins. "The Myth of the Specialized Military Contractor." *Technology Review* (April 1995): 52-58.

Kelley, Maryellen R., and Todd A. Watkins. "In From the Cold: Prospects for Conversion of the Defense Industrial Base." Science 268 (28 April 1995): 525-532.

Knezo, Genevieve J. "Defense Basic Research Priorities: Funding and Policy Issues." CCRS 90-506, Congressional Research Service, The Library of Congress, October 24, 1990.

Kornberg, Arthur. "Science in the Stationary Phase." *Science* 269 (29 September 1995): 1799.

Kettl, Donald P., and John J. Dilluio, Jr. "Cutting Government: Making Sense of Federal Reform." *The Brooking Review* (Fall 1995): 40-43.

Lawler, Andrew. "Services Target Labs to Save Money." *Science* 268 (7 April 1995): 24.

Leslie, Stuart W. *The Cold War and American Science.* New York: Columbia University Press, 1993.

Losman, Donald L., and Shu-Jan Liang. *The Promise of American Industry: An Alternate Assessment of Problems and Prospects.* Westport, CT: Quorum Books, 1990.

Lucas, Allison. "Industry and Government Share the Lab Bench." *Chemical Week* (September 28, 1994): 36.

Macilwain, Colin. "US Adopts Economy and Civilian Use as New Goals for ilitary Research." *Nature* 371 (13 October 1994): 545.

MacLachlan, Alexander. "Trusting Outsiders to Do Your Research: How Does Industry Learn to Do It?" *Research Technology Management* (November-December 1995): 48-53.

Miller, William L. "A Broader Mission for R&D." *Research Technology Management* (November-December 1995): 24-36.

Lee, Thomas H., and Proctor P. Reid, eds. *Prospering in a Global Economy: National Interests in an Age of Global Technology.* Washington DC: National Academy Press, 1991.

National Science Board. NSB 93-1, "Science and Engineering Indicators." Washington DC: U.S. Government Printing Office, 1993.

"National Security Science and Technology Strategy." Washington DC: Committee for National Security, Office of Science and Technology Policy, Executive Office of the President, 1995.

Neumann, David K. *Technology: Shaping the Face of Future Conflict*. Washington, DC: Industrial College of the Armed Forces, National Defense University, 1992.

*New World Vistas: Air and Spaec Power for the 21st Century*. Washington, DC: USAF Scientific Advisory Board, HQ USAF, December 15, 1995.

Odlyzko, Andrew M. "We Still Need Unfettered Research." *Research Technology Management* (January-February 1996): 9-11.

"Policy Statements on the Defense In-House Laboratories." Management Analysis Memorandum 66-2. Washington, DC.: Office of the Director of Defense Research and Engineering, 1 July 1966.

Rich, Ben R., and Leo Janos. *Skunk Works: A Personal Memoir of My Years at Lockheed*. New York: Little, Brown and Company, 1994.

Rifkind, Malcolm. "Global Teaming Crucial to Future." *Aviation Week and Space Technology* (May 29, 1995): 54-56.

Rosenberg, Nathan, and L.E. Birdzell, Jr. *How the West Grew Rich*. New York: Harper Collins, 1987.

Sanders, Ralph, ed. *Defense Research and Development*. Washington, DC: Industrial College of the Armed Forces, 1968.

Sanders, Ralph, ed. *Science and Technology: Vital National Resources*. Mt. Airy, MD: Lomond Books, 1975.

Schine, Eric. "Washington Won't Gallop to the Rescue." *Business Week* (January 9, 1995): 64.

Scott, William B. "Tech Transfer Impact Remains Elusive." *Aviation Week and Space Technology* (November 7, 1994): 42-44.

Scott, William B. "Technology Transfer Support Wavers." *Aviation Week and Space Technology* (October 23, 1995): 57-60.

Single, Arthur W., and William M. Spurgeon. "Creating and Commercializing Innovation Inside a Skunk Works." *Research Technology Management* (January-February 1996): 38-41.

Stafford, Thomas P. *America at the Threshold: America's Space Exploration Initiative*. Washington, DC: U.S. Government Printing Office, 1991.

Stephens, Richard H., and Eston T. White, eds. *Science and Technology*. Washington, DC: National Defense University,1983.

*Star 21: Strategic Technologies for the Army of the Twenty-First Century*. Board on Army Science and Technology, Commission on Engineering and technical Systems, National Research Council. Washington, DC: National Academy Press, 1992.

Stone, Richard. "DOD to Cut Big Bucks From Campus Research." *Science* 266 (7 October 1994): 19.

Velocci, Anthony L. "Top-Scoring Companies Pursue Similar Strategies." *Aviation Week and Space Technology* (May 29, 1995): 39.

Worden, Simon P. *SDI and the Alternatives*. Washington DC: National Defense University Press, 1991.

# ABOUT THE AUTHOR

Lieutenant Colonel Doug Beason (U.S. Air Force) currently serves as Chief, Livermore Detachment, where he is the liaison to the Department of Energy's Lawrence Livermore National Laboratory, representing the Assistant Secretary of Defense (Nuclear, Chemical and Biological Warfare). He has also worked for the President's Science Advisor on the White House staff under both the Clinton and Bush adminstrations, where he was the liaison to the National Space Council and the key White House staffer for space science and technology at NASA and DOD; as an Associate Professor of Physics and Director of Faculty Research at the USAF Academy; and as a Research Advisor to the National Research Council.

Lt. Col. Beason is a graduate of the Air Force Academy and the Industrial College of the Armed Forces, where he conducted research for this study. He received an M.S. and a Ph.D. in physics from the University of New Mexico. He is the author of several scientific articles and eleven novels and short works of fiction. His novel, *The Trinity Paradox,* was nominated for a Nebula Award, the first work of fiction to receive that honor.

*U.S. G.P.O.:1997-418-297:60017